Thermodynamics: Theory & Practice

The science of energy and power.

by D. James Benton

Copyright © 2016 by D. James Benton, all rights reserved.

Foreword

Thermodynamics is the branch of physical science that deals with energy in its various forms, including heat and work. Energy is a property of systems that can be stored and transferred to other systems. Energy is not the "ability to perform work." This is a misnomer. A system may contain considerable energy without having the capacity to perform any work. Thermodynamics is the key to understanding how things work. In this text we will explore classical (or macroscopic) as well as statistical (microscopic) thermodynamics, properties, processes, and heat engines. A variety of applications will be presented and all software is included.

This book is not intended to be a textbook on thermodynamics or to replace any of the excellent textbooks that are already available. I hope you will find this to be a helpful companion to such texts. I cover several topics (e.g., microscopic point of view, speed distributions, and probability) that are never covered in textbooks on classical (i.e., macroscopic) thermodynamics and are often reserved for graduate courses. I consider these topics to be essential to understanding the whole of thermodynamics and many practical applications in particular. I hope this presentation will inspire you to dig deeper into this fascinating field.

All of the examples contained in this book,
(as well as a lot of free programs) are available at...
https://www.dudleybenton.altervista.org/software/index.html

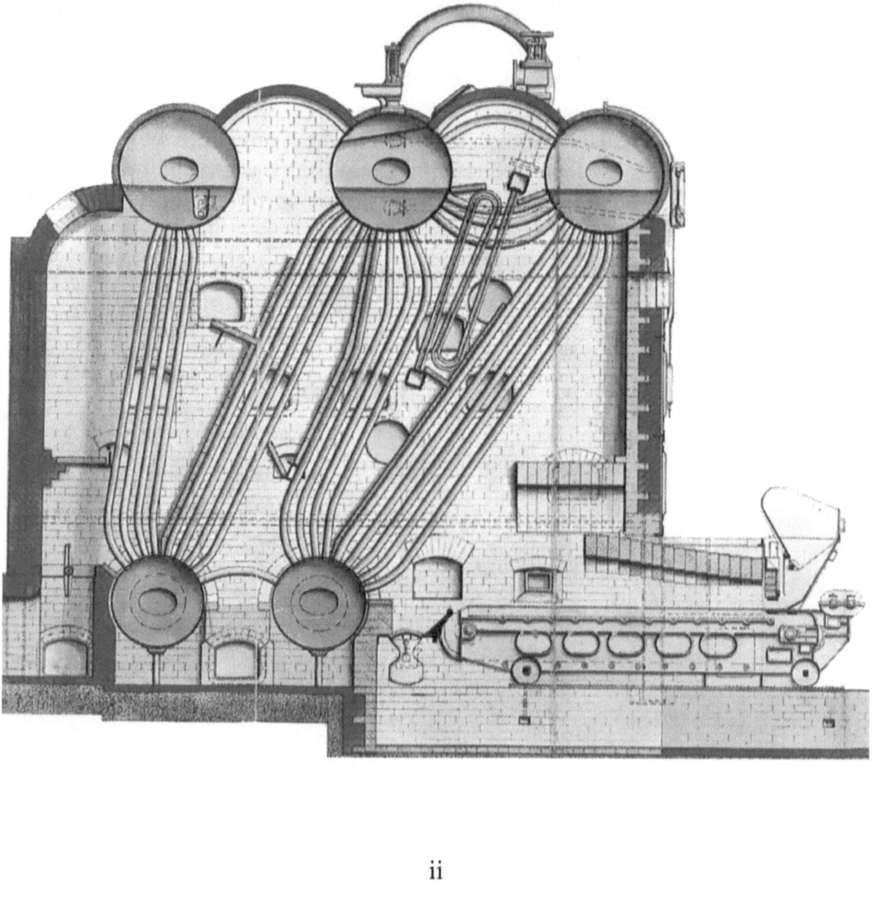

Table of Contents

page

Foreword i
Chapter 1. Systems, Heat, and Work 1
Chapter 2. The First Law of Thermodynamics 3
Chapter 3. Microscopic vs. Macroscopic 9
Chapter 4. System of Particles 15
Chapter 5. Entropy & Probability 19
Chapter 6. The Second Law of Thermodynamics 21
Chapter 7. Free Energy & Maxwell's Relations 27
Chapter 8. Equations of State 30
Chapter 9. Saturation Properties 36
Chapter 10. Specific Heats 39
Chapter 11. Developing Thermodynamic Properties 41
Chapter 12. Carnot Cycle 44
Chapter 13. Steam Turbine Test 46
Chapter 14. Regenerative Rankine Cycle 51
Chapter 15. Gas Turbine Heat Balance 54
Chapter 16. Simple Combined Cycle 59
Chapter 17. Vapor-Compression Refrigeration Cycle 62
Chapter 18. Otto Cycle 67
Chapter 19. Diesel Cycle 69
Chapter 20. Brayton Cycle 71
Chapter 21. Lenoir Cycle 73
Chapter 22. Stirling Cycle 75
Chapter 23. Ericsson Cycle 77
Chapter 24. Throttling Natural Gas 79
Appendix A: A Microscopic Perspective on Availability and Irreversibility 82
Appendix B. van der Waals EOS Program 93
Appendix C. List of Refrigerants and Properties 100
Appendix D. Approximate Properties for Ideal Gases 101

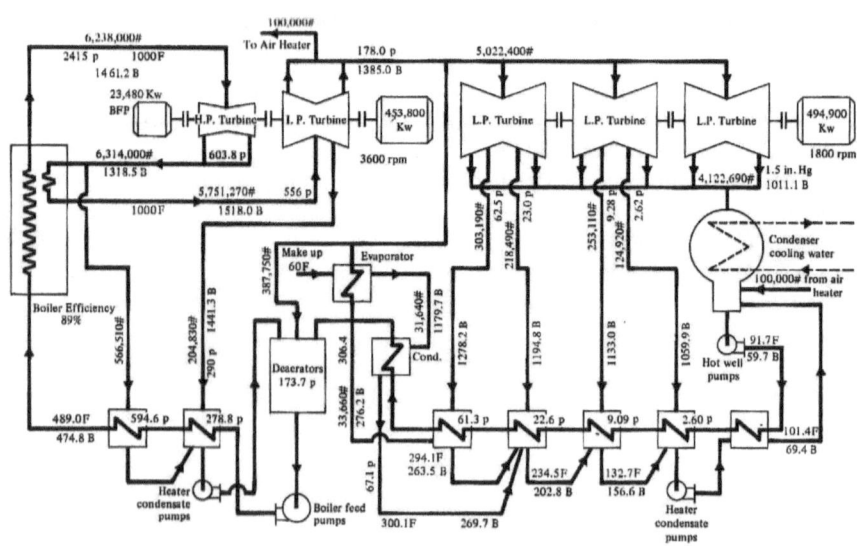

Chapter 1. Systems, Heat, and Work

After energy, systems are the most important concept in thermodynamics and are essential even to understanding energy in its various forms. A system is an abstract group of one or more objects, separated from everything else by a boundary. Although a system boundary may correspond to a physical boundary, such as a tank, this is not a requirement. A system boundary has no mass and occupies no space (i.e., has no thickness).

Furthermore, a system boundary may move, objects and energy may pass through it, and it may not be physically possible to construct a corresponding physical boundary. An example of this would be the oxygen molecules in a room. It is not possible–and may not even be desirable–to separate the oxygen molecules from the nitrogen and other gases in the room, but this doesn't preclude us from considering a system containing only the oxygen molecules.

Even in the case of something as simple as a tank, it may be advantageous to consider the contents and the tank separately. We can imagine any shape or form of system boundary and even multiple overlapping system boundaries. There are an infinite number of such selections. All selections for a system boundary are not equal. Some selections are far more useful than others. There is quite an art to the selection of a system boundary. Once we have selected a system boundary, it is essential that we handle it properly and consistently.

There are three basic types of system boundaries: 1) open, 2) closed, and 3) isolated. An open system can exchange mass and energy with its surroundings. A closed system can exchange energy, but not mass with its surroundings. An isolated system cannot exchange mass or energy with its surroundings. Closed and isolated systems are often confused and there is much useless debate over this confusion.[1]

Heat is that transient form of energy that crosses a system boundary by virtue of a temperature gradient. Work is that transient form of energy that crosses a system boundary by virtue of a force. Systems do not contain heat or work and neither can be stored. Only energy is stored. All other forms of energy that cross a system boundary apart from mass can be classified as either heat or work. For example, electricity is a form of work. Electrons flow because of a force, not a temperature. Systems do not contain temperature. Temperature– when applicable–may be a measure of the energy of a system.

Because energy is a property of objects–that is, objects can contain energy– when mass crosses a system boundary (i.e., an open system), energy is transferred from the system to its surroundings (outflow) or from the

---

[1] As we shall see, the Second Law of Thermodynamics apples equally to all three types of systems, although the equations are different. Some have mistakenly concluded that the Second Law applies only to isolated or closed systems, but not to open systems. This is nonsense.

surroundings to a system (inflow). It is essential to recognize that the surroundings include everything that isn't inside the system boundary. For convenience, the surroundings may be separated into the immediate and the distant or ultimate. The three types of systems are illustrated in the following figure:

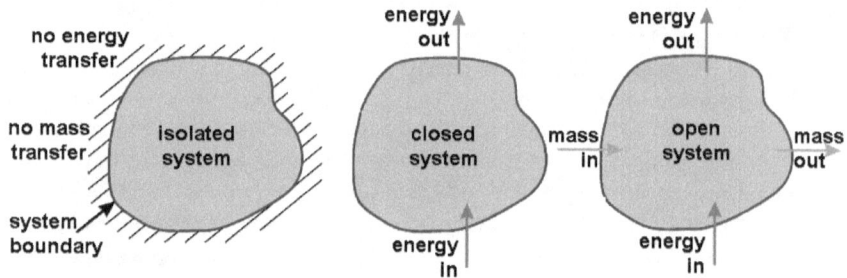

For an ideal closed system (e.g., a cylinder fitted with a frictionless piston that doesn't leak), the work is equal to the integral of force with respect to distance. For a piston, force is equal to pressure times area, so that:

$$W = \int p\,dV \qquad (1.1)$$

The work per unit mass (i.e., W/m) is given the symbol w. Specific volume is defined as volume per unit mass (i.e., V/m) and is given the symbol v. These two definitions can be substituted into Equation 1.1 to obtain:

$$w = \frac{W}{m} = \int p\frac{dV}{m} = \int p\,dv \qquad (1.2)$$

For a continuously flowing open system, power is equal to the integral of force times velocity or pressure differential times volumetric flow rate. The volumetric flow rate is equal to the specific volume times the mass flow rate; thus we have the following equation for the ideal power of a continuously flowing open system:

$$\frac{dW}{dt} = \frac{dm}{dt}\int v\,dp \qquad (1.3)$$

Chapter 2. The First Law of Thermodynamics

The First Law of Thermodynamics (1LoT) is the conservation of energy. This principle has been concluded from countless observations over many decades. In it simplest form the 1LoT can be stated: the sum of the energy in any system plus its surroundings is constant. This follows logically from the observation that energy is conserved and the definition of the surroundings of any system being everything else. The energy inside an isolated system must be constant, as it does not exchange energy with it's surroundings. This can be stated mathematically by:

$$\frac{dE_{SYS}}{dt} = \dot{E}_{SYS} = 0 \qquad (2.1)$$

A dot over a symbol is shorthand for the derivative with respect to time or the time rate of change. By convention, heat transfer into a system is considered positive and work done by a system is considered positive. The 1LoT for a closed system is then:

$$\dot{E}_{SYS} = \dot{Q} - \dot{W} \qquad (2.2)$$

The thermal energy of an object that exists by virtue of its state, independent of gross motion or chemical reaction potential, is called *internal energy* and is given the symbol U. This internal energy may be due to several factors, including molecular motion and intermolecular potential fields. The specific internal energy of an object is defined as the internal energy divided by the mass of the object and is given the symbol *u*. This same relationship of total and specific properties, along with upper and lower case symbols, is used throughout thermodynamics.

The kinetic energy of an object that exists by virtue of its gross motion is equal to the mass times the velocity squared divided by two. The specific kinetic energy is equal to this divided by the mass, or $V^2/2$.

The potential energy of an object by virtue of its elevation within a uniform gravitational field is equal to the mass times the elevation times the acceleration of gravity. The specific potential energy is equal to this divided by the mass, or *gz*.

The total energy of an object is the sum of these three components and is given the symbol $E=U+mV^2/2+mgh$. The specific total energy is equal to this divided by the mass, or $e=u+V^2/2+gz$.

Since there is no exchange of mass or energy between an isolated system and its surroundings, the total energy and mass must be constant. If such a system undergoes a change from state 1 to state 2, the 1LoT for an isolated system becomes:

$$u_1 + \frac{V_1^2}{2} + gz_1 = u_2 + \frac{V_2^2}{2} + gz_2 \qquad (2.3)$$

A closed system can exchange energy with its surroundings in the form of heat or work so that the 1LoT for a closed system becomes:

$$Q + m_1\left(u_1 + \frac{V_1^2}{2} + gz_1\right) = W + m_2\left(u_2 + \frac{V_2^2}{2} + gz_2\right) \qquad (2.4)$$

Since $m_2 = m_1$ we can also write:

$$q + u_1 + \frac{V_1^2}{2} + gz_1 = w + u_2 + \frac{V_2^2}{2} + gz_2 \qquad (2.5)$$

Here $q = Q/m$ and $w = W/m$ are the specific heat transfer and work, respectively. If mass is moving out of a system into the surroundings it does so at some pressure; thus, this process involves work. Power is the rate of doing work. The power associated with mass moving across a system boundary is equal to the pressure times the volumetric flow rate or $p \cdot dV/dt$. Specific volume is defined as volume per unit mass and is given the symbol v.[2] The volumetric flow rate is equal to the specific volume times the mass flow rate or $dV/dt = v \cdot dm/dt$. In order to account for the rate of change of energy due to this process we must add this term to the previous ones; thus we have the following relationship for energy for a flow of mass across a system boundary:

$$\frac{dm}{dt}\left(u + \frac{V^2}{2} + gz\right) + pv\frac{dm}{dt} = \dot{m}\left(u + pv + \frac{V^2}{2} + gz\right) \qquad (2.6)$$

The two terms $u + Pv$ appear so often together in thermodynamics that they are combined and given the symbol h and the name enthalpy. The 1LoT for an open system then becomes:

$$\dot{Q} + \sum_{inlets}\dot{m}_{in}\left(h + \frac{V^2}{2} + gz\right)_{in} = \frac{dE_{SYS}}{dt} + \dot{W} + \sum_{exits}\dot{m}_{out}\left(h + \frac{V^2}{2} + gz\right)_{out} \qquad (2.7)$$

The conservation of mass requires that:

$$\sum_{inlets}\dot{m}_{in} = \frac{dm_{SYS}}{dt} + \sum_{exits}\dot{m}_{out} \qquad (2.8)$$

In order to use these equations to solve problems, properties are needed. Tables of thermodynamics properties for various substances are readily available on-line. An Excel® Add-In for steam properties is provided along with the on-line archive that accompanies this book. We will discuss how to use these

---

[2] The specific volume is equal to one over the density. While density may be very useful in fluid mechanics, specific volume is much more useful in thermodynamic calculations.

tables and how to create tables of thermodynamic properties in a later chapter. For now, we will simply calculate properties using the spreadsheet functions.

A word about units... The twenty-first century engineer must be comfortable using a variety of units. If you think that the same SI units are used the world over, you've been misinformed. Even the staunchest proponents of the metric system can't agree on whether to use kPa or bars, let alone °C or °K or Kelvins. Some insist on using kg/m^3, while others insist on using gm/cm^3. English units are used through many industries in the United States and the power industry in particular. English units are even mixed with metric throughout the power industry world wide, although it's popular to refer to the former as "U.S. Customary Units." Perhaps the British have forgotten that they invented them. Who knows? After the Brexit they may start using them again. In any event, get comfortable with them all, because you'll have to work with what you're given. Don't fuss about it. There are too many people fussing already.

Example 2.1

Consider a cylinder filled with 5 pounds mass (lbm) of steam and fitted with a frictionless piston that doesn't leak.[3] A weight keeps the pressure constant at 100 pounds force (lbf) per square inch absolute (psia). The initial temperature is 500°F. Heat is added until the temperature of the steam reaches 1000°F. What is the work and heat transfer?

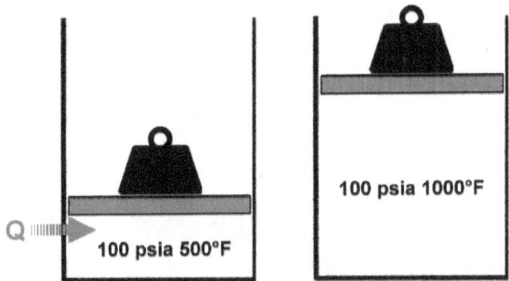

The work is equal to the force times the distance, which is equal to the pressure times the change in volume. The change in volume is equal to the mass times the change in specific volume.

$$W = \int \vec{F} \cdot d\vec{x} = \int p dV = p(V_2 - V_1) = p m(v_2 - v_1) \quad (2.9)$$

The initial and final specific volumes are 5.588 and 8.658 ft³/lbm, respectively, making the work equal to:

$$W = P m(v_2 - v_1) = \left(\frac{100 lbf}{in^2}\right)\left(\frac{12 in}{ft}\right)^2 (5 lbm)\left[(8.658 - 5.588)\frac{ft3}{lbm}\right] = 221{,}041 ft \cdot lbf \quad (2.9)$$

---

[3] This takes some imagination, because such a device couldn't be constructed.

The heat transfer is given by Equation 2.4, recognizing that the cylinder isn't moving and the change in elevation of the steam is negligible (which is not true for the weight). The initial and final values of internal energy are 1372.1 and 1175.9 British Thermal Units per pound mass (BTU/lbm)[4].

$$Q = W + m(u_2 - u_1) = \frac{(221{,}041\,ft \cdot lbf)}{\left(778.17\,\frac{ft \cdot lbf}{BTU}\right)}(5lbm)\left[(1372.1 - 1175.9)\frac{BTU}{lbm}\right] = 1265.1 BTU \quad (2.9)$$

Example 2.2

Consider a tank connected to a pipe through a valve. The tank is initially empty and everything is insulated so that there is no heat transfer.[5] Steam is flowing in the pipe at 500 kPa and 300°C. The valve is opened and the tank is filled. The volume of the tank is 0.25 m³. When the pressure in the tank reaches that in the pipe, the valve is closed. What is the final temperature of the steam in the tank? What is the mass of steam in the tank? Only consider the inside of the tank, as indicated in gray. Velocity and gravity are negligible.

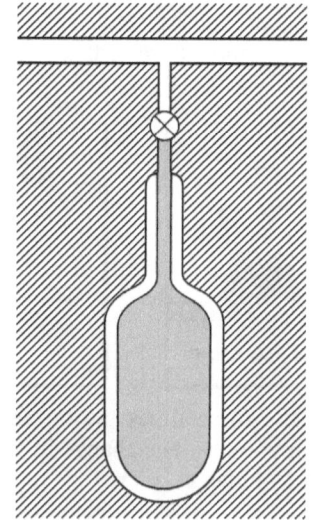

The process described in this problem is governed by Equation 2.7, integrated over time:

$$Q + m_{in}\left(h + \frac{V^2}{2} + gz\right)_{in} + m_1 e_1 = W + m_2 e_2 + \dot{m}_{out}\left(h + \frac{V^2}{2} + gz\right)_{out} \quad (2.10)$$

---

[4] Note that BTU is an acronym and is always capitalized.
[5] Hatch marks are often used to denote insulation and no heat transfer (i.e., adiabatic).

In this case $Q=0$, $W=0$, $m_f=0$, and $m_{out}=0$. The steam in the pipe may be flowing, but the steam inside the tank isn't; therefore, $e_f=u_2$, which means that the final internal energy in the tank is equal to the enthalpy in the pipe. The solution can be found using the Excel® Solver Add-In.[6]

	A	B	C	D	E
1	Example 2.2 - SI units				
2	P	T	v	h	u
3	kPa	°C	m³/kg	kJ/kg	kJ/kg
4	500	300	0.5226	3065	2803
5	500	461.5	0.6750	3402	3065
6	m	0.3704		error	0

Set E6=E5-D4, select (●)Value of: zero (0), by changing B5. The result is 461.5°C, which means the tank will get hotter. You can feel the rise in temperature after you fill a scuba tank with air. The mass is equal to the volume divided by the specific volume 0.25/0.06750=0.3704 kg.[7]

> This process is called **uniform flow, uniform state**. The state within the control volume may change over time, but at any instant it is uniform throughout the control volume. The flow rates in or out of the control volume may vary over time, but the state as mass enters or leaves the control volume is constant throughout the process.

Example 2.3

Consider a turbine into which 50 kg/s of steam enters continuously at 10 bar and 500°C and leaves at 0.1 bar and 50°C. The system shown below is insulated so that there is no heat transfer. At what rate does power leave the system boundary through the shaft? Ignore the velocity of the steam as well as any mechanical losses.

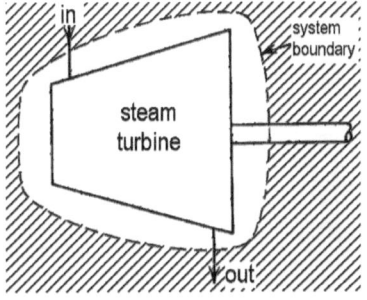

This is an application of Equation 2.7. There is only one mass flow, one entering, and one exiting enthalpy, so that the conservation of energy reduces to:

---

[6] You may need to enable this Add-In if you haven't already done so. Go to File... Options... Add-Ins... Go... find the Solver Add-In and check the box [x] next to it.
[7] The internal energy, u, must be calculated (h-pv), as the steam property Add-In doesn't provide u directly.

$$\dot{W} = \dot{m}(h_{in} - h_{out}) = \left(\frac{50 kg}{s}\right)\left[(3479 - 2592)\frac{kJ}{kg}\right] = 44{,}351 kW \qquad (2.11)$$

The Excel® steam property Add-In is used again:

	A	B	C	D
1	Example 2.3 - Metric units			
2	P	T	v	h
3	bar	°C	m³/kg	kJ/kg
4	10	500	0.3541	3479
5	0.1	50	14.87	2592
6	m	50	kg/s	
7	W	44,351	kW	

Chapter 3. Microscopic vs. Macroscopic

Up until this point we have been discussing systems (e.g., cylinders, pistons, pumps, and turbines) and materials (e.g., steam and air) on a macroscopic level. We know that all of these things are made up of much smaller objects (i.e., molecules, atoms, and subatomic particles). A view of these smaller objects will be called microscopic, although atoms are much too small to be seen under a microscope.

Consider an isolated system (i.e., an insulated box) containing a number of particles. The total energy of an isolated system is constant (Equation 2.1). The energy will be distributed among the particles. We know from observation that the energy of individual atoms is not the same and changes over time as they interact; therefore, the way in which the total energy is distributed among the particles in an isolated system changes over time, but the total remains the same.

At this point it is worth your time to watch atoms collide–at least in a simulation. Julio Gea-Banacloche is professor and head of the Physics Department at the University of Arkansas. Back in 1997 he wrote a Java applet that simulates hard spheres bouncing around in a box. His simulation uses very basic collision calculations, yet in just a few seconds the particles distribute themselves into the same range of speeds. This happens regardless of how they start out, even if initially only one particle is moving.

You can find the applet and the source code for Julio's simulation on-line, but may not be able to run it, due to security issues with Java. I have created a Windows® application that can be downloaded from my web site along with the C source code. This is what it looks like:

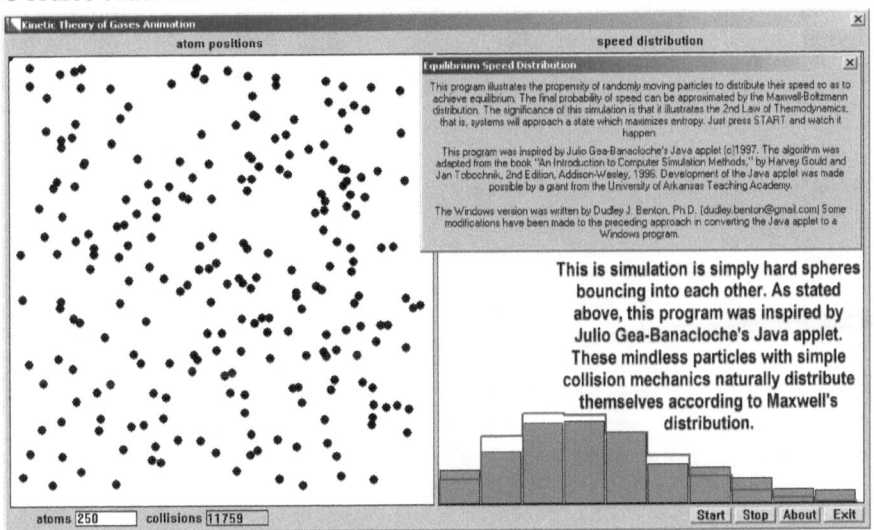

9

The current (and changing) distribution of particle speeds is shown by the red bar graph. The theoretical distribution (which we will derive next) is shown by the blue stair-step. No matter how the particles are started, they eventually approach the same distribution. Other than the blue line, the theoretical distribution doesn't appear anywhere in the source code. The particles aren't forced into this distribution by the code. Things just work out this way.

Think about it... The faster a particle is moving, the more likely it is to hit another one. The slower a particle is moving, the more likely it is to be hit by a faster one. Even if initially only a single particle is moving, it will collide with the others and start them moving, just like billiard balls. The particles don't all approach the same average speed. We also know this from billiards, as sometimes two collide and one of them stops.

In this simulation the total energy remains constant, as we have assumed this to be an isolated system. We could effectively add energy by giving one or more of the particles an extra push. We could remove energy by slowing one or more down. If we don't add or remove any particles, this is a closed system and Equation 2.2 applies.

We know from experiments that monatomic gases (e.g., helium, neon, argon, krypton, xenon, and radon) behave much like these particles. The internal energy of a monatomic gas is also known to be:

$$U = \frac{3}{2}RT\left(\frac{m}{MW}\right) \qquad (3.1)$$

where R is the ideal gas constant, T is the temperature, and MW is the molecular weight. We must use an absolute temperature scale whenever multiplying or dividing (i.e., °K or °R, never °C or °F). We also recall that:

$$R = kN_A \qquad (3.2)$$

where k is Boltzmann's[8] constant and N_A is Avogadro's number. The mass is equal to the molecular weight times the number of particles divided by Avogadro's number. The internal energy of n particles is then:

$$U = \frac{3}{2}nkT \qquad (3.3)$$

This gives us a relationship for the total energy, but not how it's distributed among the individual particles. We must return to the collisions for insight into this aspect of particle behavior. The 250 particles in the simulation were divided into ten groups by speed and the number in each group drawn as red bars. For the purposes of derivation we will assume that there are so many particles in the box that we can use differentials rather than discrete groups to represent the

---

[8] After Ludwig Eduard Boltzmann Austrian physicist (1844-1906).

particles. This is not unreasonable, considering that there are about twenty-five quadrillion atoms in a single cubic centimeter of air at room temperature.

We have already used the symbol V for volume and v for specific volume. Although we haven't mentioned it, we customarily use C for specific heat. Therefore, we will use c for speed, which fits in with *celerity*, the Latin word for *swiftness of movement*. We have already used p for pressure, so we will use P for probability. We expect the probability, P, of a particle surviving a collision and staying at its current speed can be described by some distribution function, F, so that $P=F(c)$.

For the next part of this derivation we must work in velocity space.[9][10] In physical space we have three dimensions: x, y, and z. The distance from the center, r, is equal to $sqrt(x^2+y^2+z^2)$. In velocity space we have Vx, Vy, and Vz. The speed, c, is equal to $sqrt(Vx^2+Vy^2+Vz^2)$, which is similar to the radius. A differential volume in physical space is $4\pi\, r^2$. The volume is equal to the integral of this with respect r from 0 to r, which is $4\pi\, r^3/3$. The differential volume in velocity space will be $4\pi\, c^2$ and we will integrate this from 0 to ∞.

We know the likelihood of a single particle colliding with another is proportional to its speed. The likelihood of collisions involving a population of particles is also proportional to the number of surviving particles. Therefore, the change with respect to speed of the surviving particles, *dF/dc*, is proportional to *-cF*, that is, the number of surviving particles diminishes with the product of the speed and the number of particles. This gives us the following differential equation:

$$\frac{dF}{dc} = -2\beta^2 cF \quad (3.4)$$

2β is the constant of proportionality. The square assures us that this will be a positive number and the 2 has been for convenience. Equation 3.4 is an ordinary differential equation, which can be separated:

$$\int \frac{dF}{F} = -2\beta^2 \int c\, dc \quad (3.5)$$

The solution is:

$$\ln(F) = -\beta^2 c^2 + \ln(\alpha^3) \quad (3.6)$$

---

[9] For additional discussion and an expanded version of this same derivation, I heartily recommend Felix Pierce's excellent text: Pierce, F. J., *Microscopic Thermodynamics*, International Textbook, Scranton, PA, 1968. I have used the same symbols and notation as Pierce, which should facilitate your reading of his more complete text.

[10] Another excellent text is: Sonntag, R. E., and G. J. van Wylen, *Fundamentals of Statistical Thermodynamics*, John Wiley & Sons, New York, 1966. These two texts cover many additional topics that are not the primary focus of this book.

a is the constant of integration arising from the indefinite integrals. The cube and the logarithm, $ln(\alpha^3)$, have been used for convenience. Equation 3.6 can be solved for F, the survival distribution function.

$$F = \alpha^3 e^{(-\beta^2 c^2)} \tag{3.7}$$

The differential portion of particles in a differential element of velocity space is equal to the differential volume times the survival:

$$\frac{dn}{n} = \alpha^3 e^{(-\beta^2 c^2)} 4\pi c^2 dc \tag{3.8}$$

Recall that c is the radius in this space, the area of a sphere is $4\pi r^2$, and the volume is the integral of the area with respect to the radius. The volume of a sphere is $4\pi r^3/3$. We can integrate this over all velocities and particles to determine the constant a.

$$\int_0^n dn = 4\pi n \alpha^3 \int_0^\infty e^{(-\beta^2 c^2)} c^2 dc \tag{3.9}$$

$$\alpha = \frac{\beta}{\sqrt{\pi}} \tag{3.10}$$

We now leave velocity space and return to particle space. The number of particles at any given speed is equal to:

$$f(c) = \frac{4}{\sqrt{\pi}} \beta^3 c^2 e^{(-\beta^2 c^2)} \tag{3.11}$$

We use $f(c)$ for this distribution in particle space to distinguish it from $F(c)$, which was in velocity space. We can now plot this distribution as a function.

	A	B	C	D	E	F	G	H	I	J
1	Maxwell-Boltzmann Speed Distribution									
2	cβ	f(cβ)								
3	0.0	0.0000								
4	0.1	0.0099								
5	0.2	0.0384								
6	0.3	0.0823								
7	0.4	0.1363								
8	0.5	0.1947								
9	0.6	0.2512								
10	0.7	0.3002								
11	0.8	0.3375								
12	0.9	0.3603								
13	1.0	0.3679								
14	1.1	0.3608								
15	1.2	0.3412								
16	1.3	0.3118								
17	1.4	0.2761								
18	1.5	0.2371								
19	1.6	0.1979								
20	1.7	0.1606								
21	1.8	0.1269								

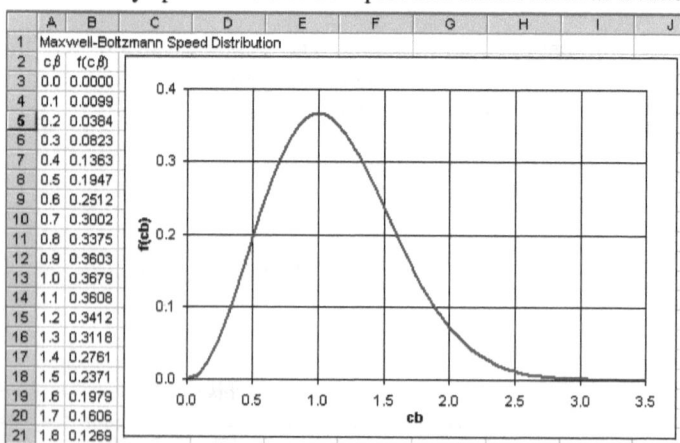

The mass of a single atom is equal to the molecular weight divided by Avogadro's number and will be given the symbol μ. The kinetic energy in a single atom is $\mu c^2/2$. The total kinetic energy of the particles is equal to the internal energy and is given by the following integral:

$$U = \frac{n\mu}{2}\int_0^\infty c^2 f(c)\,dc = \frac{2n\mu}{\sqrt{\pi}}\beta^3\int_0^\infty e^{(-\beta^2 c^2)}c^4\,dc \qquad (3.12)$$

Integration produces:

$$U = n\mu\frac{3}{2\beta^2} \qquad (3.13)$$

Combining with Equation 3.3 we get:

$$U = \frac{3}{2}nkT = \frac{3n\mu}{4\beta^2} \qquad (3.14)$$

Equation 3.14 can be solved for β:

$$\beta^2 = \frac{\mu}{2kT} \qquad (3.15)$$

We can now substitute β back into Equation 3.11 to get the velocity distribution function:

$$f(c) = \sqrt{\frac{2}{\pi}}\left(\frac{\mu}{kT}\right)^{\frac{3}{2}} c^2 e^{(-\beta^2 c^2)} \qquad (3.16)$$

The root-mean-square (rms) speed is found by substituting Equation 3.15 into 3.14 and recognizing that the average kinetic energy per particle is at this speed.

$$c_{rms} = \sqrt{\frac{3kT}{\mu}} \qquad (3.17)$$

This figure shows c_{rms} for Argon as a function of temperature.

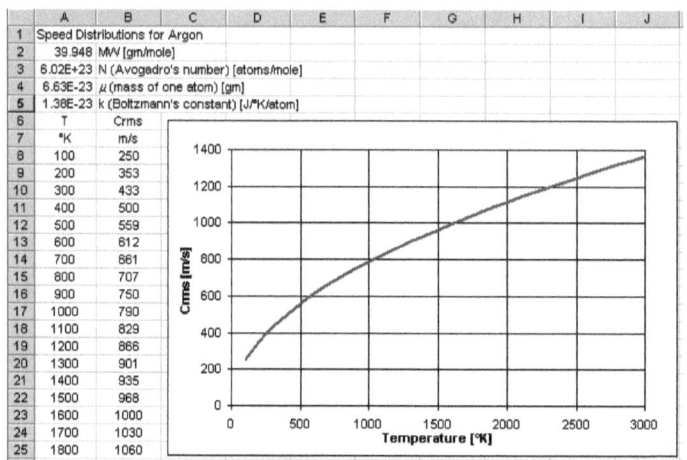

This figure shows the distribution of speeds for Argon at several temperatures:

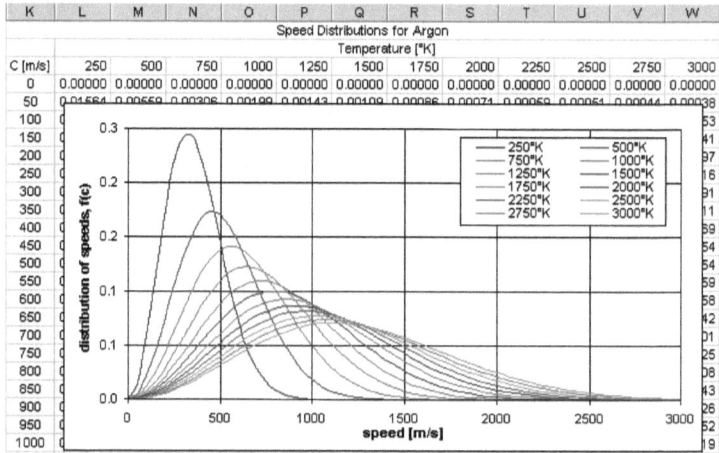

The area under the curves is the same (=1). The peak occurs at a higher speed for higher temperatures, so the height must decrease with temperature.

Chapter 4. System of Particles

In the previous chapter we considered the probability that a particle would be at a particular speed and how a population of particles will naturally distribute the speeds into groupings that exhibit a certain shape by virtue of their interactions (i.e., bouncing around). Experiments with particles (e.g., protons, neutrons, and electrons) and waves (e.g., light, X-rays, gamma rays) have shown that the energy of objects doesn't simply vary continuously, like a ball rolling down a hill. Rather, energy takes on discrete values, more like a ball on stairs.

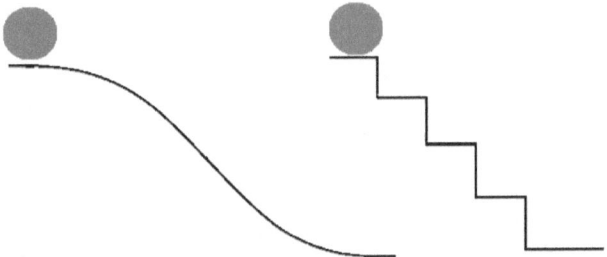

We refer to these as *energy levels*. Furthermore, we recognize that the situation isn't as simple as a ball on stairs. For instance, the same value of energy can be present, but in different forms. There are many combinations of velocity and elevation that result in the same sum: $½V^2+gz$. Rather than calling it *grouped*, we speak of energy as being *partitioned*.

We began our discussion with monatomic gases, as these are the simplest real particles. Argon atoms contain energy by virtue of their motion (i.e., bouncing around). More complex molecules can also spin and vibrate. The following figure shows the variation with temperature of several specific heats. These come from the NASA Glenn tables, an excellent source of data for thousands of substances.[11] You will find a spreadsheet in the on-line archive containing this data along with lookup functions to access it. A copy of the paper is also included.

---

[11] McBride, B. J., Zehe, M. J., and Gordon, S., "NASA Glenn Coefficients for Calculating Thermodynamic Properties of Individual Species," NASA TP-20020211556, 2002.

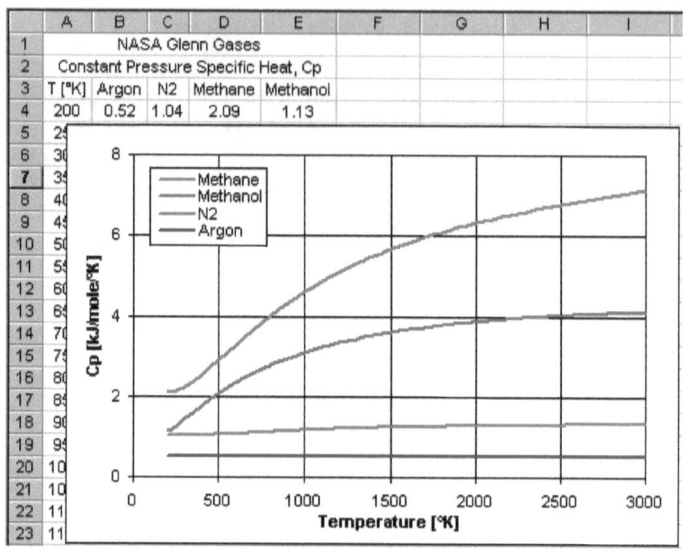

Specific heat is a measure of a substance's capacity to store internal (i.e., thermal) energy. The argon curve is flat, because the atoms don't do much besides bounce around. Nitrogen can spin and vibrate, so its curve changes a little with temperature. Methanol is lopsided and can spin and vibrate in several different ways. Methane molecules have a surprising number ways of moving about so as to store thermal energy. These specific heats change with temperature because the various modes of storing thermal energy are excited at different temperatures.

We recognize that the energy in a system of particles is partitioned into several levels. We will use e_i to represent the energy in level i and n_i to represent the number of particles in level i. The number of particles must add up to N and the energy must add up to E.

$$N = \sum n_i \qquad (4.2)$$

$$E = \sum n_i e_i \qquad (4.3)$$

The following figure represents the energy levels and the particles in each level:

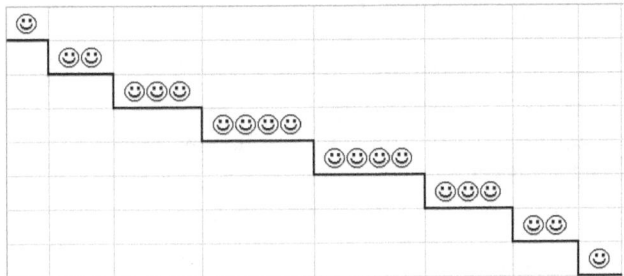

We will now consider two different ways of changing the energy of this system. The first way would be to change the number of particles on each step. This would be analogous to the last figure in the preceding chapter that showed how the speed distribution changed with temperature. Such a change is illustrated in the following figure:

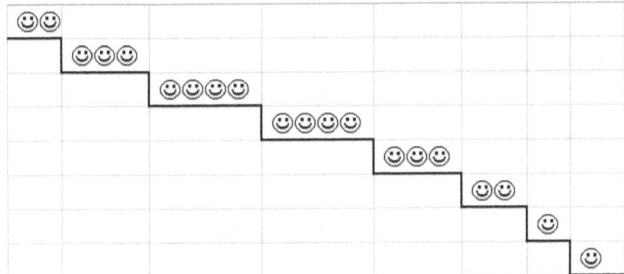

This would be a change in n_i. The other type of change would be in e_i. In this second change the number of particles on each step remains the same, but the height of the steps changes:

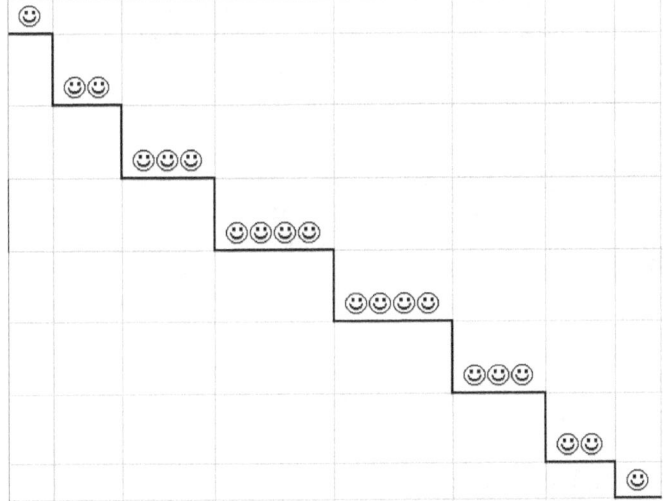

Mathematically we have:

$$\Delta E = \sum e_i \Delta n_i + \sum n_i \Delta e_i \quad (4.4)$$

If we do not add or remove any particles from the system it will be *closed*. We can adapt Equation 2.4 (i.e., the 1LoT for a closed system) to accommodate our particles and energy levels.

$$\Delta Q - \Delta W = \Delta E = \sum e_i \Delta n_i + \sum n_i \Delta e_i \quad (4.5)$$

We know for this closed system that the most work we could extract from it is equal to *pdV*, which we could only be realized with a frictionless piston. Upon substitution and rearrangement, Equation 4.5 becomes:

$$\Delta Q = p\Delta V + \sum e_i \Delta n_i + \sum n_i \Delta e_i \quad (4.6)$$

Everything on the right side of this equation is a property of the system; therefore, the left side must also be a property. We know that adding or removing heat will change the temperature of the system, but there is more to this than temperature alone. We will introduce another property, call it *entropy*, and give it the symbol S. For convenience we will define S such that:

$$\Delta Q = T\Delta S \quad (4.7)$$

Equation 4.5 becomes:

$$T\Delta S - p\Delta V = \Delta E = \sum e_i \Delta n_i + \sum n_i \Delta e_i \quad (4.8)$$

We recall that changing the temperature directly impacts how many particles are at each speed, regardless of whether or not any work is done.[12] This means that the first term on the left side of Equation 4.8 is linked to the first term on the right side. The second terms must also be linked. This means that heat is related to the distribution of particles among the energy levels and work is related to the energy levels themselves. For a closed system at rest, the thermal energy is equal to U; thus for a macroscopic system we can deduce the following relationship on a per mass basis:

$$Tds - pdv = du \quad (4.9)$$

There is no way to directly measure entropy, but we can measure p, v, u, and T, so that we can calculate s. As the first term on the left side of Equation 4.8 (TΔS) is related to the first term on the right ($\Sigma e_i \Delta n_i$) and the number of particles in each energy level is related to probabilities, we expect that entropy will also be related to probabilities.

---

[12] If the size of the box doesn't change ΔV is zero and there is no work, regardless of whether p changes or not.

Chapter 5. Entropy & Probability

Consider a coin in a box. It could be either heads or tails. The probability, P, is 1/2. If there were two coins, the probability of them both being heads (or tails) would be 1/4 and so on for more coins. Consider the following box containing 4 coins, separated by a partition:

If the partition were removed we would get the following:

Recall the definition of a system and its boundary from Chapter 1. A system is an abstract grouping. A system boundary may or may not correspond to a physical one. We can conceptualize as many system boundaries (i.e., groupings) as we want. No matter how many ways we choose to group these four coins, the probability of them all being heads is the same. There is one and only one mathematical relationship that produces this same result no matter how we choose to group the coins: the logarithm.

$$\ln\left(\frac{1}{4}\right) + \ln\left(\frac{1}{4}\right) = \ln\left(\frac{1}{16}\right) = \ln\left(\frac{1}{2}\right) + \ln\left(\frac{1}{8}\right) = \ln\left(\frac{1}{2}\right) + \ln\left(\frac{1}{2}\right) + \ln\left(\frac{1}{2}\right) + \ln\left(\frac{1}{2}\right) \quad (5.1)$$

This combined probability is a characterization of the entire system and not the likelihood of a single coin toss. It's clearly not the same as the probability that a particle will have a particular speed, which we derived before. There is a conceptual as well as mathematical gap in our description of thermodynamic systems: proving the equation, "*s=k ln (W)*." Here *W* replaces *P* and is said to be the number of ways in which a system could *realize* (in the sense of *attain*) a particular state. A Google search on s=klnw will direct you to these three sites:

http://hyperphysics.phy-astr.gsu.edu/hbase/therm/entrop2.html
https://en.wikipedia.org/wiki/Boltzmann%27s_entropy_formula
http://www.eoht.info/page/S+%3D+k+ln+W

The first site shows a pyramid of dice and follows a similar line as the coins. It presents the equation, but doesn't say where it comes from. The second site discusses the equation and how it relates to pressure and permutations, but doesn't derive it. The third site cuts straight to the point: Ludwig Boltzmann came up with it, but could never rigorously prove it. Albert Einstein even criticized Boltzmann for this on several occasions. It would seem that no one has yet been able to rigorously prove it. If Boltzmann couldn't prove it and Einstein accused him of hand waving, you should expect no more from me ☺

From the preceding chapter we have:

$$T\Delta S = \sum e_i \Delta n_i = n \sum e_i \frac{\Delta n_i}{n} \qquad (5.2)$$

We know that e_i is proportional to kT and that $\Delta n_i/n$ is proportional to $\Delta P/P$, where P is the probability. Provided that n is very large and ΔS is very small, we can write:

$$TdS = nkT \frac{dP}{P} \qquad (5.3)$$

The solution of Equation 5.3 is:

$$\frac{S}{n} = s = k \ln(P) \qquad (5.4)$$

In spite of not being able to rigorously prove the steps between Equations 5.2 and 5.3, the result works quite well and has withstood the test of time since Boltzmann introduced it in 1901. This equation is engraved on his tombstone and the critics have long since quit harping on the lack of rigor. Not only does it have the right units and magnitude, it meets the expectation of being the logarithm of a probability.

Chapter 6. The Second Law of Thermodynamics

The Second Law of Thermodynamics (2LoT) is not so obvious as the First. Still, it has been concluded from countless observations over many decades. Presenting the 2LoT in a way that can be easily understood has been a passion of the Author's for forty years, even before publishing the paper included here as Appendix A.

In science—which rests firmly upon the foundation of experiment—we must rely on inductive reasoning: beginning with particular observations and discerning the general principles responsible for what we see. Based on observations we formulate a hypothesis. Experiments are devised and performed to test the hypothesis. If necessary, the hypothesis is revised and more experiments conducted until the matter is settled to the satisfaction of the proverbial skeptic. So it is with the 2LoT.

It has long been observed that some systems are capable of doing useful work while others seem unable. We suspect that there is some quality (i.e., property) of a system in addition to energy that has not yet been captured in the 1LoT. We further observe that a change of state or disposition of a system can alter this yet undefined property. Consider the following system:

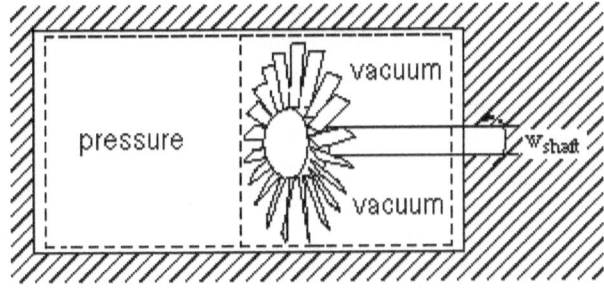

There is a membrane (i.e., dotted line) running down the middle of this box, separating it into two compartments, one pressurized with air and the other evacuated. The box is insulated so that there will be no heat transfer. Furthermore, the seal around the shaft does not leak so that no mass enters or leaves.

We instinctively know that when the membrane is ruptured, the air will quickly flow from one side to the other until the pressure is balanced throughout. When this happens the fan will turn and work could be extracted from the shaft. From the 1LoT we know that, if work is extracted from this system, the energy of the air must diminish; thus, we expect it to get cooler. By rupturing the membrane we have not only extracted work, we have caused something to get spontaneously cooler.

Two questions come to mind: "What has changed?" and "How much work could we extract from this process?" We also instinctively know that this process cannot be *undone*. Even if we were to heat the air back to its original temperature, this won't turn the fan. We hypothesize that something about the air has permanently changed and that this change has something to do with being separated and contained on one side of the membrane.

In order to *undo* this process, we would have to return the system to its original state. We couldn't repair a membrane, but we could replace it with a valve and open that to let the air rush out and fill the entire box. We could also replace the fan with a pump. We could close the valve, pump the air back over to the left side, and then heat (or cool) it back to its original temperature. After having done so, we could claim to have *reversed* the process.

We could do this over and over again, as long as the final state of the system is returned to the original. We could also measure the difference between the initial and final states of the system. We could conduct any number of related experiments, for instance, boiling and condensing steam in a cylinder and allowing this to move a weight.

We hypothesize that there is some maximum amount of work that could be extracted and that this is a property of the system. We will call this the *reversible work*. We will call the difference between the reversible work and the actual work *irreversibility* and give it the symbol *I*. It is easy to conduct an experiment from which we extract no work at all (e.g., leave out the fan). In that case the irreversibility is equal to the reversible work, as the actual work is zero.

We have already considered the work of displacing a frictionless piston and have noted that this would be the maximum. A rigid box without a shaft would produce no work, which would be the minimum. These two processes correspond to the reversible work and the irreversibility. Since the reversible work is the maximum, the irreversibility must always be greater than or equal to zero. For an incremental process involving a closed system we have by definition:[13]

$$pdV - \delta W = \delta I \geq 0 \tag{6.1}$$

In this equation the symbol δ is used to denote an inexact differential, as compared to the symbol d, which indicates an exact differential. Pressure and volume are properties of a system, so it doesn't matter how the system gets from p_1 to p_2 or from V_1 to V_2, only that it has arrived at the second state. The same is true for elevations.

Work and heat are *path* functions, which means it matters how you get from point 1 to point 2. When you're driving from Denver to Las Vegas, it matters whether you take the northern route on I-70 or the southern route through Albuquerque to I-40. Either way you will go from 5000' to 2000'.

---

[13] This is the definition of the irreversibility.

From the 1LoT we have:

$$\delta Q = \delta W + dU \tag{6.2}$$

Multiplying Equation 4.9 times the mass and substituting it into 6.1 and 6.2 yields:

$$TdS - \delta Q = \delta I \geq 0 \tag{6.3}$$

Equation 6.3 is often written:

$$dS > \frac{\delta Q}{T} \tag{6.4}$$

which applies to a closed system for which there is no work. A reversible process is one in which there is no irreversibility (i.e., $\delta I=0$), in which case:

$$dS = \left(\frac{\delta Q}{T}\right)_{rev} \tag{6.5}$$

Because a system is an abstract grouping, we could just as easily select the surroundings as the system and the system as the surroundings. Equation 6.4 applies to any closed system with a rigid boundary (i.e., no work). More specifically, it applies to the system and the surroundings. We can add the change in entropy of the system and the surroundings and conclude the following:

$$dS_{uni} = dS_{sys} + dS_{sur} = \delta Q \left(\frac{1}{T_{sys}} - \frac{1}{T_{sur}}\right) \geq 0 \tag{6.6}$$

Heat only flows from the hotter to the colder. If $\delta Q>0$ (i.e., into the system), then T_{sur} must be greater than T_{sys} and $(1/T_{sys}-1/T_{sur})>0$, so the inequality holds. If $\delta Q<0$ (i.e., out of the system), then T_{sur} must be less than T_{sys} and $(1/T_{sys}-1/T_{sur})<0$, so the inequality holds in either case.

For an open system we must also consider the entropy entering and leaving the system through the boundary:

$$\frac{dS_{sys}}{dt} + \sum_{exits} \dot{m}_{out} s_{out} - \sum_{inlets} \dot{m}_{in} s_{in} \geq \frac{\dot{Q}}{T} \tag{6.7}$$

Adding the system and the surroundings as before:

$$\frac{dS_{uni}}{dt} = \frac{dS_{sys}}{dt} + \sum_{exits} \dot{m}_{out} s_{out} - \sum_{inlets} \dot{m}_{in} s_{in} - \frac{\dot{Q}}{T} \geq 0 \tag{6.8}$$

Whether we're considering an isolated, closed, or open system, the resulting change in entropy for the system plus the surroundings is always greater than or equal to zero. This is the basis for the statement that entropy is always increasing. There is a clear direction (i.e., timeline) for all thermodynamic processes.

It is very important to note here that, while the derivation of Boltzmann's equation, s=k·ln(P), may be lacking in rigor, it was not used in this chapter. The 2LoT for isolated, closed, and open macroscopic systems does not depend on this relationship being strictly true for those systems. The derivation of the 2LoT as presented here is rigorous.

Example 6.1

A cup (250 ml) of coffee begins at the ideal brewing temperature (96°C or 369.15°K) and cools to room temperature (23°C or 296.15°K). What is the change in entropy for this process? You may consider coffee to have the same thermal properties as water and ignore the thermal energy in the cup as well as any evaporation.

The mass is equal to:

$$m = \frac{250ml}{\left(\frac{10^6 ml}{m^3}\right)\left(0.00104 \frac{m^3}{kg}\right)} = 0.240kg \tag{6.8}$$

The heat transfer is given by Equation 2.4:

$$Q = (0.240kg)\left(96.5 - 402.1 \frac{kJ}{kg}\right) = -73.5kJ \tag{6.9}$$

The change in entropy for the system (i.e., the coffee) plus the surroundings (i.e., the room) is given by Equation 6.6:

$$dS_{uni} = (-73.5kJ)\left(\frac{1}{369.15°K} - \frac{1}{296.15°K}\right) = 0.049 \frac{kJ}{°K} \tag{6.10}$$

Example 6.2

A mug (500 ml) of beer begins at the ideal serving temperature (4°C or 227.15°K) and warms to room temperature (23°C or 296.15°K). What is the change in entropy for this process? You may consider beer to have the same thermal properties as water and ignore the thermal energy in the mug as well as any evaporation.

The mass is equal to:

$$m = \frac{500ml}{\left(\frac{10^6 ml}{m^3}\right)\left(0.00100 \frac{m^3}{kg}\right)} = 0.500kg \tag{6.8}$$

The heat transfer is given by Equation 2.4:

$$Q = (0.500kg)\left(96.5 - 16.8 \frac{kJ}{kg}\right) = 39.8kJ \tag{6.9}$$

The change in entropy for the system (i.e., the beer) plus the surroundings (i.e., the room) is given by Equation 6.6:

$$dS_{uni} = (39.8kJ)\left(\frac{1}{277.15°K} - \frac{1}{296.15°K}\right) = 0.0092 \frac{kJ}{°K} \qquad (6.10)$$

We see from these two examples whether cooling or heating, there is always an increase in entropy.

Example 6.3

What is the efficiency of the turbine in Example 2.3 and what is the rate of entropy production? Assume there is no heat transfer.

Steam turbines are much hotter (500°C) than their surroundings (e.g., 30°C); therefore, any heat transfer would be out of the turbine to the surroundings. This would result in less energy available for doing work, so the best scenario would be no heat transfer (i.e., an *adiabatic* process). The maximum work (or power) would be for a reversible process. Equation 6.5 tells us for a reversible process, if δQ=0, then dS=0. The maximum power would result if the exit entropy were the same as the inlet (i.e., an *isentropic* process).[14]

The inlet enthalpy is 3479 kJ/kg and the entropy is 7.764 kJ/kg/°K. The enthalpy at the exit pressure and inlet entropy is 2461 kJ/kg. The enthalpy at the exit pressure and temperature is 2592. The maximum power is equal to:[15]

$$\dot{W} = \dot{m}(h_1 - h_{2S}) = \left(\frac{50kg}{s}\right)\left[(3479 - 2461)\frac{kJ}{kg}\right] = 50,895 kW \qquad (6.11)$$

The actual power from before was 44,351 kW, making the efficiency equal to 87.1%, which is typical for a steam turbine. Equation 6.8 gives us the rate of entropy generation:

$$\frac{dS_{uni}}{dt} = \left(\frac{50kg}{s}\right)\left[(8.174 - 7.764)\frac{kJ}{kg°K}\right] = 20.51 \frac{kW}{°K} \qquad (6.12)$$

---

[14] This calculation is used so frequently that the Excel® steam property has a function to return enthalpy as a function of pressure and entropy as well as temperature as a function of pressure and enthalpy. If you didn't have this useful function but did have entropy as a function of pressure and temperature, you could use the Solver Add-In to find the solution.

[15] It is customary to use the subscripts 1 and 2 to indicate inlet and exit respectively and 2S to indicate the exit condition at constant entropy.

The spreadsheet calculations are shown below:

	A	B	C	D	E
1	Example 6.3 - Metric units				
2	P	T	v	h	s
3	bar	°C	m³/kg	kJ/kg	kJ/kg/°K
4	10	500	0.3541	3479	7.764
5	0.1	46	14.67	2461	7.764
6	0.1	50	14.87	2592	8.174
7	m=	50	kg/s		
8	Wtmax=	50,895	kW		
9	Wmax=	44,351	kW		
10	eff=	87.1%			
11	dS/dt=	20.51			

Chapter 7. Free Energy & Maxwell's Relations

Combining Equations 4.9, 6.2, and 6.5 we find that the most energy that could possibly be converted to work at constant volume (*isochoric*) and temperature (*isothermal*) on a per mass basis is given by:

$$a = u - Ts \qquad (7.1)$$

This useful combination of properties is called the *Helmholtz free energy* and is given the symbol **a**.[16] The maximum work for a constant pressure (*isobaric*) process is given by:

$$g = h - Ts \qquad (7.2)$$

This useful combination of properties is called the *Gibbs free energy* and given the symbol **g**.[17] Thermodynamic properties are interrelated in remarkable ways. We will start by taking the derivative of the Helmholtz free energy:

$$da = du - Tds - sdT \qquad (7.3)$$

Equation 7.3 can be combined with 4.9 (**Tds-pdv=du**) to form:

$$da = -sdT - pdv \qquad (7.4)$$

The property a is uniquely specified by T and v so that by the chain rule of calculus we can write:[18]

$$da = \left(\frac{\partial a}{\partial T}\right)_v dT + \left(\frac{\partial a}{\partial v}\right)_T dv \qquad (7.5)$$

Comparing the terms in Equations 7.4 and 7.5 we see that:

$$\left(\frac{\partial a}{\partial T}\right)_v = -s \qquad (7.6)$$

$$\left(\frac{\partial a}{\partial v}\right)_T = -p \qquad (7.7)$$

We can go through these same steps with the Gibbs free energy.

$$dg = dh - Tds - sdT \qquad (7.8)$$

---

[16] This property is named after Hermann von Helmholtz, a German physicist (1821-1894), and is usually denoted by the letter **a** (from the German "arbeit" or work). Helmholtz is responsible for many significant developments in physics, mechanics, engineering, physiology, and psychology.

[17] This property is named after Josiah Willard Gibbs (1839-1903), an American scientist who made important theoretical contributions to physics, chemistry, and mathematics. His work in the area of thermodynamics transformed physical chemistry into a rigorous science.

[18] This particular application is known as Schwarz's theorem or Young's theorem and follows from the symmetry of second derivatives.

Next, we note that **h=u+pv**, therefore:
$$dh = du + pdv + vdp \tag{7.8}$$
which yields...
$$dh = Tds + vdp \tag{7.9}$$

so that...
$$dg = -sdT + vdp \tag{7.10}$$
as before...

$$dg = \left(\frac{\partial g}{\partial T}\right)_p dT + \left(\frac{\partial g}{\partial p}\right)_T dp \tag{7.11}$$

Comparing the terms in Equations 7.10 and 7.11 we see that:

$$\left(\frac{\partial g}{\partial T}\right)_p = -s \tag{7.12}$$

$$\left(\frac{\partial g}{\partial p}\right)_T = v \tag{7.13}$$

Returning to Equation 4.9 (**Tds-pdv=du**) and recognizing that **u** is uniquely specified by **v** and **s**:

$$du = \left(\frac{\partial u}{\partial s}\right)_v ds + \left(\frac{\partial u}{\partial v}\right)_s dv \tag{7.14}$$

Comparing the terms in Equations 4.9 and 7.12 we see that:

$$\left(\frac{\partial u}{\partial s}\right)_v = T \tag{7.15}$$

$$\left(\frac{\partial u}{\partial v}\right)_s = -p \tag{7.16}$$

Similarly, for **h** we can write:

$$dh = \left(\frac{\partial u}{\partial s}\right)_p ds + \left(\frac{\partial h}{\partial p}\right)_s dp \tag{7.17}$$

Comparing the terms in Equations 7.9 and 7.17 we see that:

$$\left(\frac{\partial h}{\partial s}\right)_p = T \tag{7.18}$$

$$\left(\frac{\partial h}{\partial p}\right)_s = v \tag{7.19}$$

Equations 7.6, 7.7, 7.12, 7.13, 7.15, 7.16, 7.18, and 7.19 can be combined to form:

$$T = \left(\frac{\partial h}{\partial s}\right)_p = \left(\frac{\partial u}{\partial s}\right)_v \quad (7.20)$$

$$p = -\left(\frac{\partial u}{\partial v}\right)_s = -\left(\frac{\partial a}{\partial v}\right)_T \quad (7.21)$$

$$v = \left(\frac{\partial h}{\partial p}\right)_s = \left(\frac{\partial g}{\partial p}\right)_T \quad (7.22)$$

$$s = -\left(\frac{\partial a}{\partial T}\right)_v = -\left(\frac{\partial g}{\partial T}\right)_p \quad (7.23)$$

Equations 7.20 through 7.23 are the first order relationships between the various thermodynamic properties. Taking the partial derivatives of these equations reveal more important relationships. The first of these is found by taking the partial derivative of the first and third terms in Equation 7.20 with respect to v at constant s and the partial derivative of the first and second terms in Equation 7.21 with respect to s at constant v and so on:

$$\left(\frac{\partial T}{\partial v}\right)_s = -\left(\frac{\partial p}{\partial s}\right)_v = \frac{\partial^2 u}{\partial s \partial v} \quad (7.24)$$

$$\left(\frac{\partial T}{\partial p}\right)_s = \left(\frac{\partial v}{\partial s}\right)_p = \frac{\partial^2 h}{\partial s \partial p} \quad (7.25)$$

$$\left(\frac{\partial s}{\partial v}\right)_T = \left(\frac{\partial p}{\partial T}\right)_v = -\frac{\partial^2 a}{\partial T \partial v} \quad (7.26)$$

$$-\left(\frac{\partial s}{\partial p}\right)_T = \left(\frac{\partial v}{\partial T}\right)_p = \frac{\partial^2 g}{\partial T \partial v} \quad (7.27)$$

These last four are called Maxwell's relations.[19] These eight equations are essential to the development of thermodynamic properties, as they relate the properties we can't measure directly to the ones we can.

---

[19] After James Clerk Maxwell Scottish scientist (1831-1879).

Chapter 8. Equations of State

All pure substances that are chemically stable (i.e., the molecules don't break down, chain together, or otherwise become something else) exhibit certain behaviors over some range of temperatures and pressures. This includes transitioning from a solid to a liquid to a vapor. While the conditions over which this behavior is exhibited may vary considerably (e.g., orders of magnitude difference in pressures), there is a certain similarity (i.e., while the scale may vary, the shape is similar).

One of the most fascinating aspects of this behavior is the manifestation of a critical point. At one specific combination of temperature and pressure, the liquid and vapor are indistinguishable. For water this occurs at a temperature of 374.15°C (705.4°F) and a pressure of 22,064 kPa (3203.6 psia). There is also a point when the solid, liquid, and vapor can all three coexist. This is called the triple point. For water this occurs at 0.01°C (32.018°F) and 0.6112 kPa (0.08866 psia).

There are many different graphs that can be drawn to represent thermodynamic properties. Perhaps the most common is pressure vs. specific volume showing lines of constant temperature. The following figure is for water:

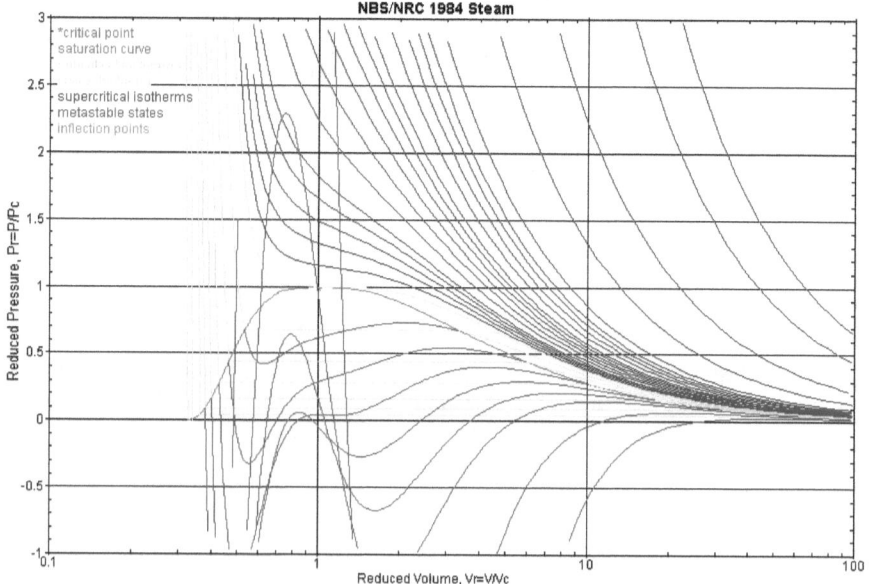

The X-axis is the specific volume divided by the specific volume at the critical point and the Y-axis is the pressure divided by the pressure at the critical point. The cyan curve is the critical isotherm. The green curves are isotherms at temperatures below the critical and the blue curves are the isotherms at temperatures above the critical.

The red curve is the locus of all pure liquid and pure vapor phases. It is also the pressures and temperatures at which liquid and vapor can coexist in equilibrium.[20] This is called the saturation line (i.e., the saturation temperature, pressure, pure liquid specific volume, and pure vapor specific volume).[21] Because of its shape the red curve is often called the *steam dome* or *vapor dome* when referring to something other than water.

The brown curves are called metastable states. These are not in equilibrium and last only a very brief time while the system is transitioning from one equilibrium state to another. The most common observance of this is boiling water on the stove. At tiny imperfections in the surface and for a very short time before a bubble can form, the liquid water can be more than 100°C (212°F). As soon as the bubble forms the temperature rushes back to 100°C (212°F). This is a real effect and can be measured.[22] The following figure is a three-dimensional representation of this phase space.

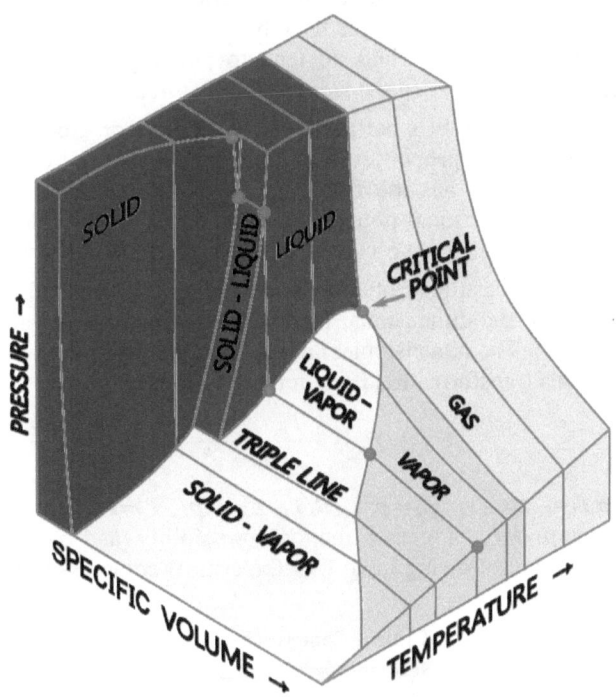

---

[20] Equilibrium describes a state that is stable and constant over an indefinite period of time (i.e., none of the thermodynamic properties are changing with time).
[21] The term saturation defines a condition in which a mixture of vapor and liquid can coexist at a given temperature and pressure.
[22] The author has done so personally on many occasions with very sensitive instruments.

As thermodynamic properties are essential in all sorts of calculations, it is no wonder that much effort has been devoted to developing equations to represent the aforementioned behavior. These are called equations of state (EOS) and there are many. The oldest and simplest was presented by van der Waals in 1873 and has the following form:

$$p = \frac{RT}{v-b} - \frac{a}{v^2} \tag{8.1}$$

The constants a and b are typically selected to obtain a best-fit data for the substance in question. Tables of such constants are readily available on-line. The first term on the right side of this equation is the same as the ideal gas EOS (i.e., $p=RT/v$) except with $v-b$ in the numerator. This term is added to account for the fact that the molecules do take up some space and it is known that the pressure of liquids increases very rapidly with decreasing density.

The second term on the right side of Equation 8.1 is an attractive force (it has a minus sign in front of it). This term has been added in recognition of the fact that the molecules of a fluid do tend to pull together if expanded very rapidly. This attraction is called the *van der Waals force*.

While this simple EOS is not particularly accurate or useful in developing tables of thermodynamic properties, it is quite remarkable in that it predicts the correct behavior (i.e., it has the right shape). Most importantly, the van der Waals EOS exhibits a critical point, liquid and vapor states below the critical pressure, and asymptotically approaches the ideal gas EOS at low density.

It is helpful to recast this Equation 8.1 entirely of dimensionless parameters by dividing each of the dimensional parameters by the critical value (indicated by the subscript C). The dimensionless parameters are given the subscript, R, for *reduced*. With this transformation Equation 8.1 becomes:

$$z_C p_R = \frac{T_R}{v_R - B} - \frac{A}{v_R^2} \tag{8.2}$$

where $A=a/RT_C v_C$, $B=b/v_C$, $Z_C = p_C v_C/RT_C$, $p_R=p/p_C$, $T_R=T/T_C$, and $v_R=v/v_C$. The combination, $z=pv/RT$, is called the compressibility and z_C is the critical compressibility. The following table lists the critical compressibility for several common substances:

Critical Compressibilities	
Substance	Zc
Water	0.230
Ammonia	0.242
Carbon Dioxide	0.275
Nitrogen	0.287
Helium	0.291
Hydrogen	0.307
van der Waals	0.375

Water has one of the smallest values and hydrogen has one of the largest. van der Waals equation predicts a critical compressibility of 3/8, which is higher than any known substance. This is the most often cited shortcoming of this EOS. An inspection of the two previous figures reveals that the first and second derivatives of the pressure with respect to volume are zero at the critical point.

The van der Waals EOS is one of a class often called *cubic* equations of state, because they can be rearranged to form a cubic equation in either specific volume or compressibility and then solved analytically for three roots. The van der Waals EOS can be recast as follows:

$$Z^3 - \left(\frac{P_R}{8T_R}+1\right)Z^2 + \left(\frac{27P_R}{64T_R^2}\right)Z - \frac{27P_R^2}{512T_R^3} = 0 \tag{8.3}$$

In order to be consistent with the observed behavior of fluids (which the van der Waals EOS is in this respect) Equation 8.3 must have three distinct real roots for $T<T_C$ (red 1 2 3), three equal real roots at $T=T_C$ (magenta 123), and only one real root (dark cyan 1) plus two imaginary roots for $T>T_C$. These three conditions are illustrated in the following figure:

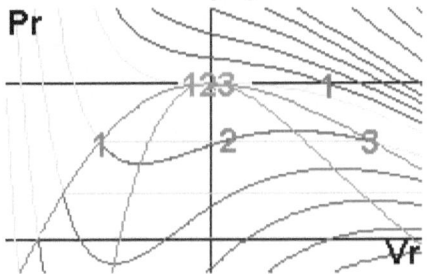

There are many EOSs, each with strengths and weaknesses. None are perfect. Some exhibit good agreement with a few properties and poor agreement with others. For example, most empirical EOSs have more than three real roots for $T<T_C$. I have written a Windows® program that will draw graphs for a dozen different EOSs and a dozen graphs for each, including the four most common empirical EOSs for water. It's called EOSPLT and can be downloaded from my web site. I have also provided a program in Appendix B that includes all of these calculations and will be discussed in Chapter 11.

As you might notice from the two previous figures, the various curves have discontinuous slopes (i.e., abrupt changes in slope). These are accurate representations of the actual behavior of real substances. All of the property curves have discontinuous slopes... except two: enthalpy, h, and entropy, s.[23] An h vs. s graph is called a *Mollier* diagram:

---

[23] A graph of internal energy, u, vs. entropy, s, would also work, but there is very little use for such a graph and I have never seen one.

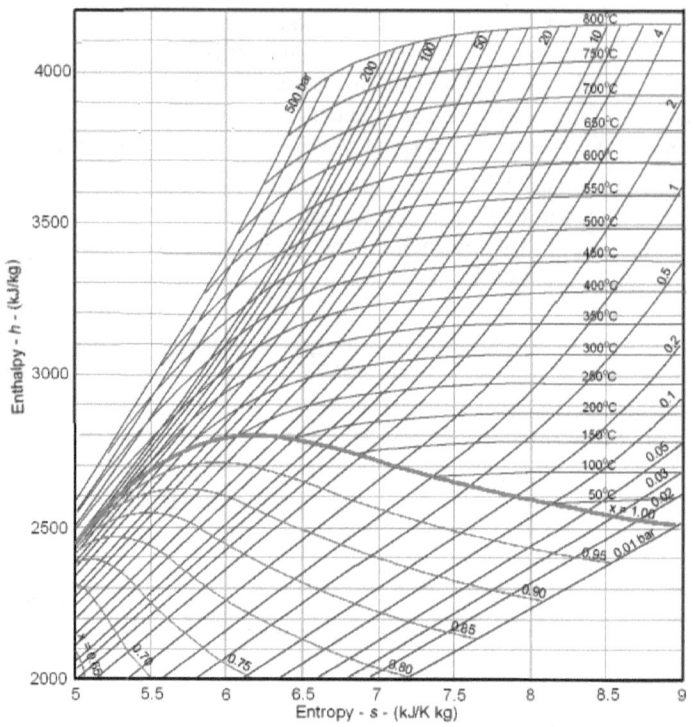

This next figure is a dimensionless Mollier diagram for water based on the same property formulation as the first figure in this chapter (i.e., the NBS/NRC 1984):

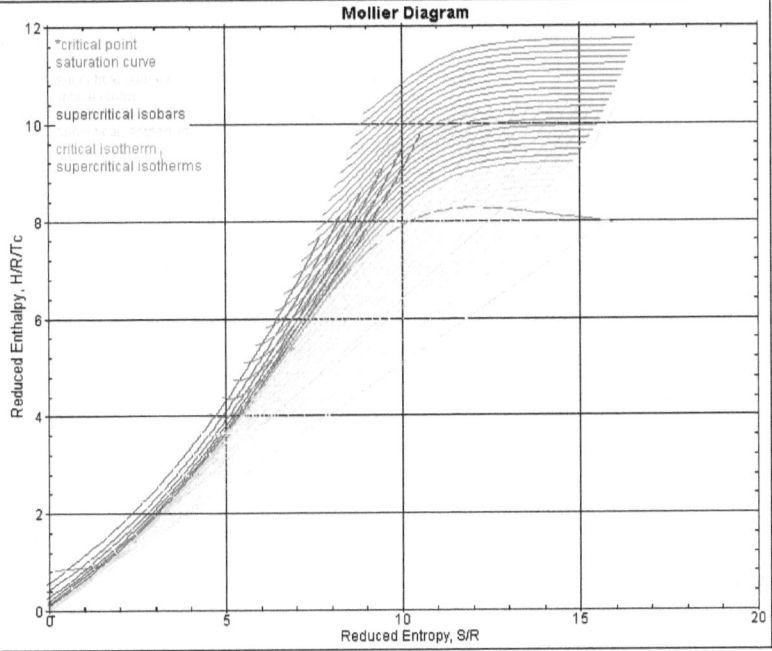

Chapter 9. Saturation Properties

Tables of thermodynamic properties typically include the saturated liquid and vapor phases, denoted by the subscripts f and g, respectively along with the compressed liquid and superheated vapor. For convenience, many tables of thermodynamic properties also include the difference between the saturated vapor and saturated liquid and denote this difference by the subscript fg.[24]

When a mixture of saturated liquid and vapor is present, the vapor fraction is given the symbol x and is called *quality*. This term came from the early days of steam power, before superheated steam was available and *dry* steam (100% quality) was preferable to *wet* steam (less than 100% quality). The saturated liquid state corresponds to $x=0$ and the saturated vapor state corresponds to $x=1$. We have the following relationships for the properties of mixtures of saturated liquid and vapor:

$$v = v_f + x v_{fg} \qquad (9.1)$$

$$u = u_f + x u_{fg} \qquad (9.2)$$

$$h = h_f + x h_{fg} \qquad (9.3)$$

$$s = s_f + x s_{fg} \qquad (9.4)$$

As we mentioned in Chapter 7, Maxwell's relations are central to the development of thermodynamic properties. We begin with Equation 7.20, which tells us that the change in enthalpy with respect to entropy at constant pressure is equal to temperature. A transition from saturated liquid ($x=0$) to saturated vapor ($x=1$) at constant pressure is also at constant temperature, T_S, the saturation temperature, which means that:

$$h_{fg} = T_S s_{fg} \qquad (9.5)$$

The Gibbs free energy across this same isotherm for any value of quality, x, is equal to:

$$g = h - Ts = h_f + x h_{fg} - T_S \left(s_f + x s_{fg} \right) \qquad (9.6)$$

Factoring this equation with respect to x yields:

$$g = h_f - T_S s_f + x \left(h_{fg} - T_S s_{fg} \right) \qquad (9.7)$$

From Equation 9.5 we see that the last term in Equation 9.7 is zero, which means that, regardless of x, g is the same. In other words, the Gibbs free energy of the saturated liquid and saturated vapor is the same. This is known as *Maxwell's Criteria*.

We will next consider the integration of Equation 4.9 along this same line from the saturated liquid to vapor.

---

[24] I have no idea why h_g-h_f is called h_{fg} instead of h_{gf}.

$$\int_{s_f}^{s_g} T_s ds - \int_{v_f}^{v_g} pdv = \int_{u_f}^{u_g} du \qquad (9.8)$$

We recognize that T_S is a constant and rearrange terms to get:

$$\int_{v_f}^{v_g} pdv = T_s s_{fg} - u_{fg} \qquad (9.9)$$

We recall that $h_{fg}=T_s s_{fg}$ and $u_{fg}=h_{fg}-p_s v_{fg}$, which yields upon substitution:

$$\int_{v_f}^{v_g} pdv = p_s v_{fg} \qquad (9.10)$$

Equation 9.10 is sometime also called Maxwell's Criteria. It has a very important implication to something we can't directly measure. Recall the brown curves in the first figure in Chapter 8. These were called metastable states that last for only a short time. The following is an expanded part of this same graph:

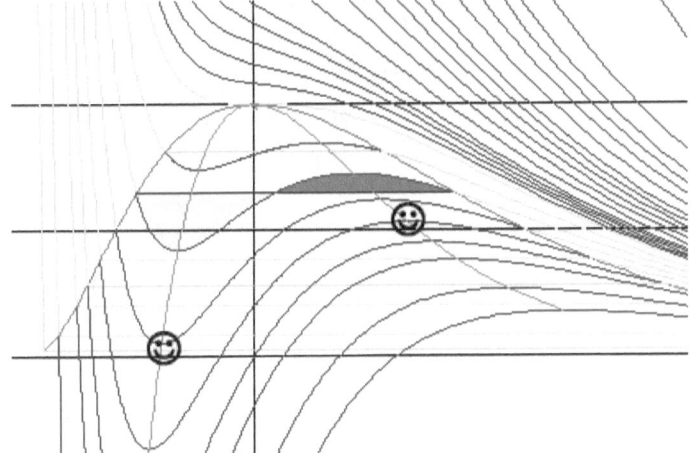

The horizontal purple line between the yellow and red shaded areas is for some constant saturation temperature and pressure. The area under this line is a rectangle and equal to the right side of Equation 9.10. The left side of Equation 9.10 is the integral along the brown curve that runs below and above the purple line. In order for this equation to hold true, the yellow and red shaded areas must be equal. They don't look equal in this figure because this is a log scale.

What this means is that once you select an equation of state (e.g., van der Waals) you have also specified the relationship between the saturation pressure and temperature. Conversely, if you have measured this relationship for some substance and seek to develop an equation of state, it must be able to satisfy this requirement. This requirement provides another step toward developing thermodynamic properties for any substance.

In this figure, the magenta curve passes through the minimum point on each brown curve (metastable states) on the left side of the critical point (the intersection of the red, magenta, and cyan curves) and the maximum on each brown curve on the right side. This locus of points is called the *spinodal curve*. This curve represents the most unstable conditions: superheated liquid on the left of the critical point or subcooled vapor on the right. This curve is important in the analysis of boiling on the one side and condensation on the other.

One of Maxwell's relations is quite useful in relating the properties at saturation. The change from saturated liquid to saturated vapor at constant pressure is also at constant temperature; Equation 7.26 can be written:

$$\frac{dp_{sat}}{dT_{sat}} = \frac{s_{fg}}{v_{fg}} \qquad (9.11)$$

Equation 9.5 can be substituted into Equation 9.11 to obtain:

$$\frac{dp_{sat}}{dT_{sat}} = \frac{h_{fg}}{T_{sat} v_{fg}} \qquad (9.12)$$

Equation 9.12 is called the Clapeyron equation.[25]

---

[25] After Benoît Paul Émile Clapeyron French engineer and physicist (1799-1864).

Chapter 10. Specific Heats

I expect the reader is already somewhat familiar with specific heats and alluded to this in Chapter 3. There are two that we will consider in more detail: constant pressure and constant volume. These are given the symbols C_P and C_V, respectively, and defined by the following equations.

$$C_P = \left.\frac{\partial h}{\partial T}\right)_{p=const} \quad (10.1)$$

$$C_V = \left.\frac{\partial u}{\partial T}\right)_{v=const} \quad (10.2)$$

These properties of a substance are relatively easy to measure and tables of values are readily available for countless substances. Some readers may recall that Equation 10.1 was the original basis for two important units of heat: the British Thermal Unit (BTU) and the calorie. The BTU was originally defined as the energy required to raise one pound of water one degree (Fahrenheit) and the calorie as the energy required to raise one gram of water one degree (Centigrade).[26] Both have been redefined in terms of joules.

As tables of single values of specific heats are common, the presumption is often made that these are constant or vary only slightly with temperature. As we have already seen in Chapter 4, the specific heat of gases can vary considerably. The same is true for liquids, as this next figure illustrates:

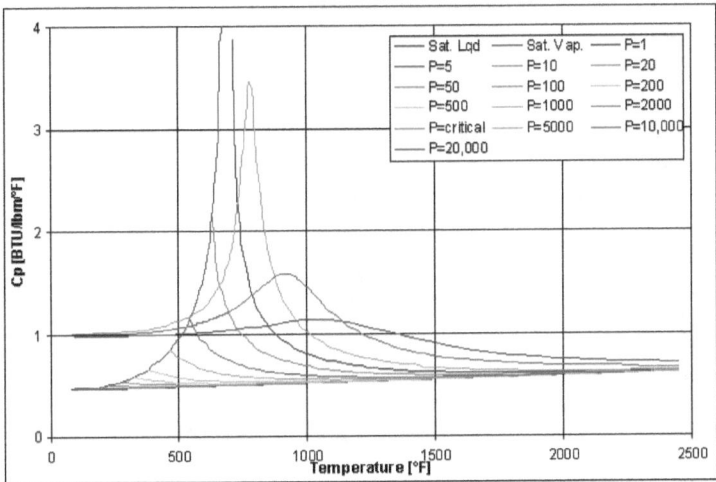

---

[26] Note there are two types of calories: gram and kilogram. These are often denoted by lower and upper case C, respectively. The ones you see on packages of food are kilogram calories. Why doesn't the FDA make this distinction clear? Because it looks bad enough to say a burger has 1200 Calories. To inform customers that a burger has 1,200,000 calories would bring on a heart attach much sooner and drastically impact sales.

39

In fact, the specific heat of the saturated liquid and vapor are infinite (or undefined), as there is a large change h or u (i.e., h_{fg} and u_{fg}) with no change in temperature (i.e., $T=T_{sat}$). These cases are of little practical importance, as we simply use the tabulated or calculated values of enthalpy and internal energy. The most useful specific heats are at the other extreme: extremely low pressure. These are designated by adding a subscript 0.

$$C_{P0} = \left(\frac{\partial h}{\partial T}\right)_{p=0} \tag{10.3}$$

$$C_{V0} = \left(\frac{\partial u}{\partial T}\right)_{v=0} \tag{10.4}$$

Before we close this chapter on specific heats we need to develop an additional relationship. Beginning with Equation 7.1 we consider a change in temperature at constant pressure:

$$\left(\frac{\partial a}{\partial T}\right)_v = \left(\frac{\partial u}{\partial T}\right)_v - s - T\left(\frac{\partial s}{\partial T}\right)_v \tag{10.5}$$

From Equation 7.23 we have:

$$s = -\left(\frac{\partial a}{\partial T}\right)_v \tag{10.6}$$

Equation 10.6 can be substituted into Equation 10.5, yielding:

$$\left(\frac{\partial a}{\partial T}\right)_v = \left(\frac{\partial u}{\partial T}\right)_v + \left(\frac{\partial a}{\partial T}\right)_v - T\left(\frac{\partial s}{\partial T}\right)_v \tag{10.7}$$

The second term on the right side of Equation 10.7 cancels out the left side. The first term on the right side is equal to C_V (i.e., Equation 10.2). Equation 10.6 can be substituted into the last term on the right side to obtain the following:

$$C_v = -T\left(\frac{\partial^2 a}{\partial T^2}\right)_v \tag{10.8}$$

Chapter 11. Developing Thermodynamic Properties

Tables or functions that provide accurate thermodynamic properties of various substances (most importantly steam) are vital to countless industrial processes and have been essential to the development of modern technology. As we have alluded to at several points, the relationships between the various properties are the key to developing such properties. Reviewing these relationships reveals the following important fact: all the other properties can be easily and unambiguously derived from the Helmholtz free energy. If we had an accurate approximation for this, we would know all of the other properties.

$$p = -\left(\frac{\partial a}{\partial v}\right)_T \tag{11.1}$$

$$s = -\left(\frac{\partial a}{\partial T}\right)_v \tag{11.2}$$

$$u = a - Ts \tag{11.3}$$

$$h = u + pv \tag{11.4}$$

From Equation 11.1 and 11.2 we see that an approximation for a as a function of T and v is needed so that we can calculate these two derivatives. By examining Equations 7.21 ($p=-\partial a/\partial v$) and 10.8 ($C_V=-T\partial^2 a/\partial T^2$) we see that only two types of data are necessary to develop thermodynamic properties: C_{V0} and $p(T,v)$. To obtain the Helmholtz free energy at any we can first integrate C_{V0}/T twice from any reference temperature along the zero isobar (i.e., at a constant pressure of zero, which corresponds to an infinite specific volume or a zero density). Second, we integrate pdv along the isotherm (i.e., at constant temperature). It is customary to set the free energy to zero for the liquid at the triple point, which is adding a constant. This two-part integration can be written:

$$a = -\left(\int pdv\right)_{T=const} - \left(\iint \frac{C_{V0}dT}{T}\right)_{p=0} + const. \tag{11.5}$$

Many equations of state can be integrated analytically, for instance the van der Waals EOS. The first term in Equation 11.5 can be obtained by integrating Equation 8.2 to arrive at Equation 11.6. For convenience, the integral can be multiplied by the critical compressibility and divided by the critical pressure and specific volume to obtain the dimensionless form:

$$\frac{Z_C}{P_C V_C}\left(\int pdv\right)_{T=const} = T_R \ln(V_R - B) + \frac{A}{V_R} + const. \tag{11.6}$$

For a constant specific heat, the second term in Equation 11.5 is simply:

$$\left(\iint \frac{C_{V0}dT}{T}\right)_{p=0} = C_{V0}[T\ln(T) - T] + const. \tag{11.7}$$

Equations 11.1 through 11.7 have been implemented in the program listed in Appendix B, which can be used to produce the following figure:

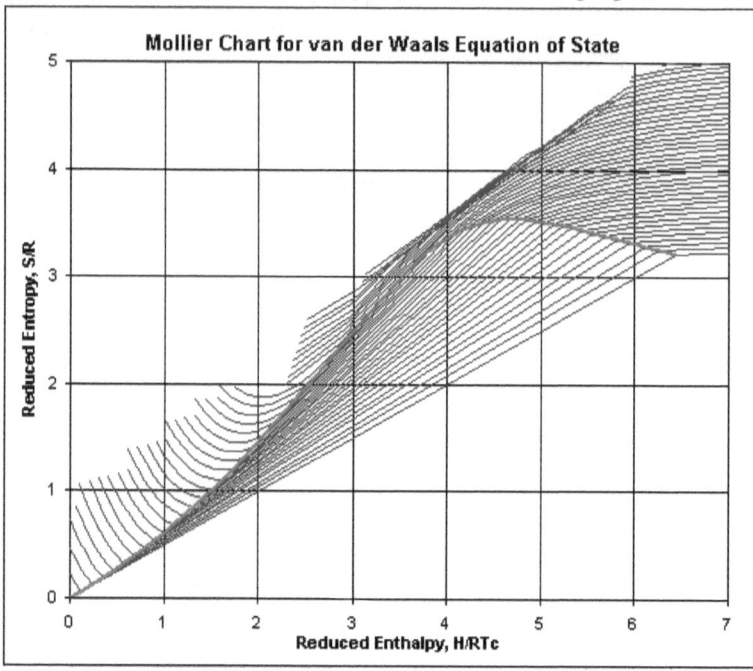

The Redlich-Kwong EOS[27] is a significant improvement over the van der Waals, which was further improved by Soave.[28] In dimensional form the RKS becomes:

$$P = \frac{RT}{(v-b)} - \frac{\alpha a}{v(v+b)} \qquad (11.8)$$

$$\alpha = \left[1 + \left(0.48508 + 1.155171\omega - 0.151613\omega^2\right)\left(1 - \sqrt{T_R}\right)\right]^2 \qquad (11.9)$$

$$a = 0.42748\left(\frac{R^2 T_C^2}{P_C}\right) \qquad (11.10)$$

$$b = 0.08664\left(\frac{RT_C}{P_C}\right) \qquad (11.11)$$

The parameter, ω, in Equation 11.9 is the Pitzer acentric factor, tables of which are readily available.[29] The following substitutions can be made to

---

[27] Redlich O. and Kwong, J. N., "On the Thermodynamics of Solutions," *Chemical Review*, Vol. 44, No. 1, pp. 233-244, 1949.
[28] Soave, G. "Equilibrium Constants from a Modified Redlich-Kwong Equation of State," *Chemical Engineering Science*, Vol. 27, No. 6, pp. 1197-1203, 1972.

facilitate transformation of the RKS into the form of Equation 8.3 for compressibility:

$$A = \frac{\alpha a}{R^2 T^2} \quad (11.12)$$

$$B = \frac{bP}{RT} \quad (11.13)$$

$$Z^3 - Z^2 + (A - B - B^2)Z - AB = 0 \quad (11.14)$$

The RKS can also be integrated analytically and used to create tables and graphs of thermodynamic properties. The program to implement this (rks.c) is included in the on-line archive and can be used to produce the following figure:

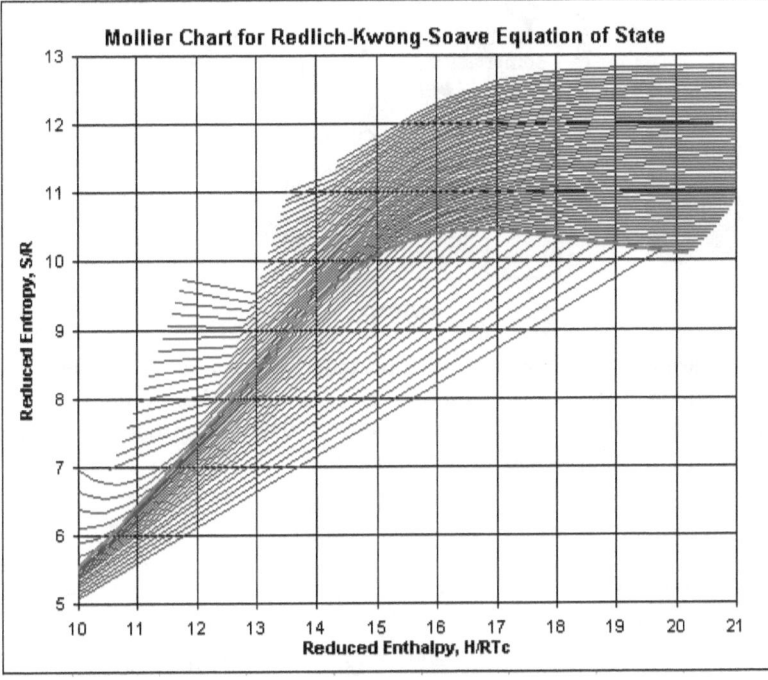

Analytical equations of state are approximations of real fluids and each will produce somewhat different graphs, as is the case with the van der Waals and RKS Mollier charts. As mentioned previously, EOSPLT will display these for you. Accurate representation of real substances requires more complicated empirical equations of state.

---

[29] Pitzer, K. S. "The Volumetric and Thermodynamic Properties of Fluids," *Journal of the American Chemical Society*, Vol. 77, No. 65, pp. 3427–3434, 1955.

Chapter 12. Carnot Cycle

The simplest thermodynamic cycle is called the *Carnot Cycle*.[30] It is the basis of comparison for all other thermodynamic cycles. The importance of entropy to the 2LoT was presented in Chapter 6. In light of this relationship, it is often informative to plot thermodynamic cycles on a temperature-entropy diagram. This is especially true of the Carnot Cycle, which is illustrated in the following figure:

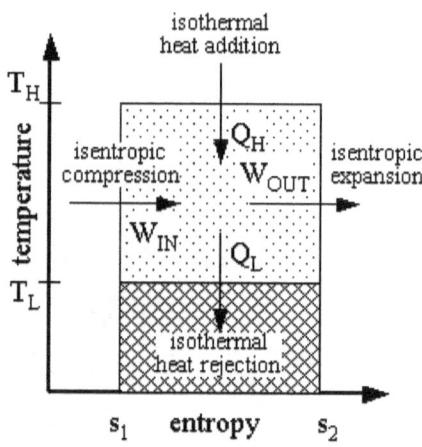

The Carnot Cycle consists of four ideal processes: 1) isothermal (i.e., constant temperature) heat addition, 2) isentropic (i.e., constant entropy) reversible expansion, 3) isothermal heat rejection, and isentropic reversible compression. From Equation 6.5, we conclude that:

$$Q_H = T_H(s_2 - s_1) \quad (12.1)$$

$$Q_L = T_L(s_2 - s_1) \quad (12.2)$$

As all four processes are reversible, the net work must equal the net heat, which means that:

$$W_{NET} = W_{OUT} - W_{IN} = Q_H - Q_L = (T_H - T_L)(s_2 - s_1) \quad (12.3)$$

The net work is equal to the area of the upper (dotted) rectangle in the preceding figure. The work out is equal to the area of both rectangles and the work in is equal to the area of the bottom (hatched) rectangle. The efficiency is equal to the work out divided by the heat in, or:

$$\eta = \frac{(T_H - T_C)}{T_H} \quad (12.4)$$

---

[30] After Nicolas Léonard Sadi Carnot (1796–1832), a French military engineer and physicist, often described as the "father of thermodynamics".

This is called the *Carnot Efficiency*. Recalling the discussion of maximum (reversible) work and heat transfer in Chapter 6, we conclude that the greatest efficiency any cycle could possibly achieve is equal to the area of the upper rectangle divided by the area of both rectangles. From the geometry of this figure, we also conclude that the most efficient cycle will have a rectangular shape. Consider the following figure of some arbitrary cycle:

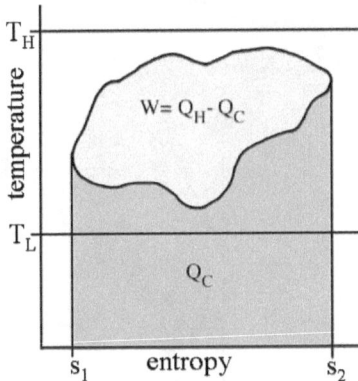

The blob at the top will always have less area than a rectangle filling the same region bounded by T_H above and T_L below, decreasing the efficiency numerator. If anything, the darker shaded area at the bottom will also have more area than a rectangle, increasing the efficiency denominator. Thus, the Carnot Cycle, with its rectangular shape, is the most efficient thermodynamic cycle possible. This has far-reaching implications. For example, a power plant operating between a flame temperature of T_H rejecting heat to the atmosphere or a body of water at T_L can never achieve an efficiency any higher than that given by Equation 12.4.

> *The Carnot Cycle is the most efficient thermodynamic cycle and the Carnot Efficiency is the most that any heat engine can possibly achieve.*

The Carnot Cycle is primarily useful as a comparison to other cycles as the optimum. It is also important in the historical development of thermodynamic cycle analyses. However, it is not possible to construct a machine that will function as prescribed by the Carnot Cycle. Some small and very impractical machines have been built to illustrate the concept, but these are not used industrially.

Chapter 13. Steam Turbine Test

A steam turbine test is a common thermodynamic problem. The efficiency of a steam turbine is determined by performing a test in accordance with American Society of Mechanical Engineers (ASME) Power Test Code (PTC) 6 (Steam Turbines) or 6.2 (Steam Turbines in Combined Cycles). Data from an actual test will be used for this example. The following figure shows a typical steam turbine, in this case for a combined cycle plant:

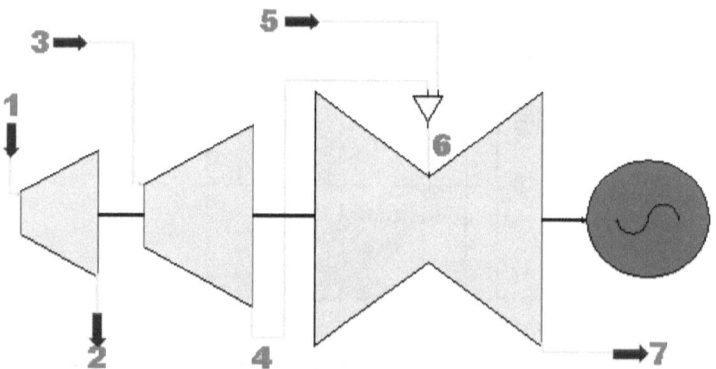

The steam turbine is divided into three sections on a common shaft that drives a single generator. The first section on the left is the high-pressure turbine (HPT). The section in the middle is the intermediate-pressure turbine (IPT) or reheat turbine. The last section on the right is the low-pressure turbine (LPT). The numbers in this figure refer to: 1) the main steam, 2) the cold reheat, 3) the hot reheat, 4) the cross-over, 5) the LP admission, 6) the combined low pressure steam, and 7) the LP exhaust.

There is a reheater plus some additional intermediate-pressure steam between points 2 and 3. The steam at point 5 comes from the low-pressure superheater. The LPT exhausts into the condenser. The flows at points 1, 3, and 5 are calculated from feed water flows, as direct measurement of vapor flow under these conditions is much less accurate.

The temperature and pressure are measured at points 1, 2, 3, and 5, which provide the state. The temperature at point 4 cannot be accurately measured. The state at point 6 can be calculated from that at points 4 and 5 along with the flows. The steam at point 7 is *wet*. That is, it is some mixture of liquid and vapor and cannot be directly measured. The velocity and moisture vary with position and there is no instrument capable of making such measurements, even if deployed at multiple positions across the flow.

There are long-standing arguments in the industry over how to measure the pressure at point 7 or that such is possible. Even the pressures at points 4 and 6 are often disputed. As the steam at point 7 is wet, the pressure and temperature are not independent. Some argue in favor of measuring the pressure and calculating the temperature, while others argue for measuring the temperature and calculating the pressure. The former is known to be problematic and the latter requires knowledge of (or enforced ignorance of) the subcooling of the condensate in the condenser.

The performance of such a steam turbine is, therefore, indeterminate when relying on steam measurements alone. This conundrum can only be solved by considering additional information. First, the generator output can be measured very accurately. The generator losses are a small portion of the total output and can be measured with acceptable accuracy. For instance, the mechanical (i.e., rotational) losses and the I^2R losses (often called *copper* losses) make up the majority of the total losses and are very repeatable. Measurement of the generator output and power factor, combined with the losses, yields the shaft power. The losses for this particular generator at four values of power factor[31] are shown in the following figure:

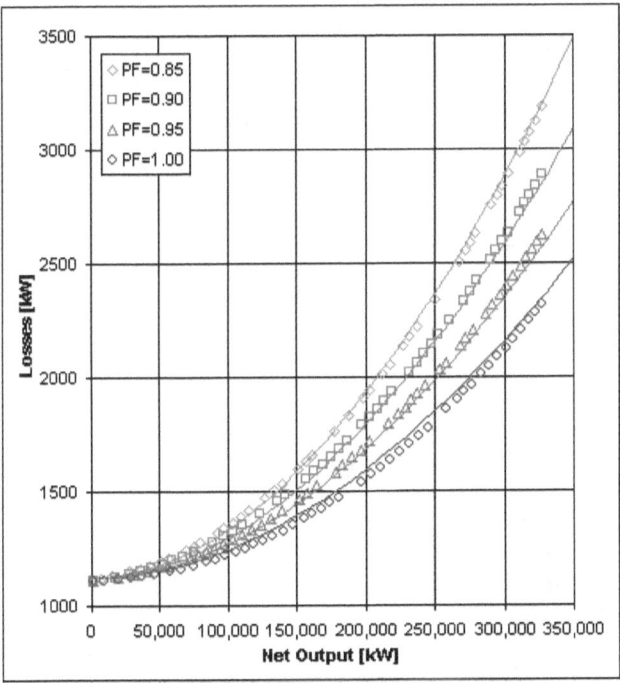

---

[31] Power factor is the ratio of the real power to the apparent power in an alternating circuit. It is equal to the cosine of the angle between the voltage and the current. "ELI the ICE man" is a way to remember that voltage (given the symbol E) leads current (given the symbol I) in an inductor (given the symbol L) and current leads voltage in a capacitor (given the symbol C).

Ken Cotton (General Electric's steam turbine guru for decades) came up with a method for correlating performance data for steam turbines exhausting into a condenser. His method involves what he called *exhaust loss*.[32] For the purposes of this example, we will assume that the method is sound and that the *exhaust loss curve* provided by the manufacturer is accurate.

This brings us to the subject of expansion lines, a method of analysis and representation employed extensively by Ken Cotton. The simplest way to illustrate this concept is to plot these points on a *Mollier Chart* (i.e., an enthalpy-entropy diagram).

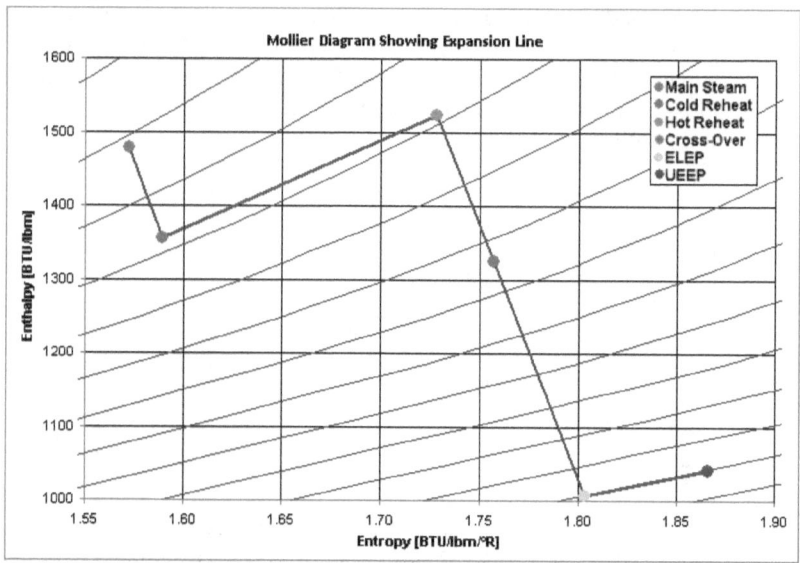

In the figure above, the thin violet curves are isobars and the thick blue segments are the expansion lines. The points are in order, starting from the top left and continuing on to the bottom right. We will first consider the last two, which are labeled ELEP (Expansion Line End Point) and UEEP (Ultimate Energy End Point). The UEEP is the actual total specific energy (in the form of enthalpy) that enters the condenser. This is the enthalpy that must be used to calculate the LPT section gross (or shaft) power.

The ELEP doesn't physically exist and can't be measured. It is a fictitious state. Recall Equations 2.7 and 2.10. The total specific energy is equal to the enthalpy plus the kinetic energy term, $V^2/2$. The velocity inside the steam turbine can't be measured and an appropriate average can't be determined. The

---

[32] Spencer, R. C., Cotton, K. C., and Cannon, C. N.,"A Method for Predicting the Performance of Steam Turbine-Generators 16,500 kW and Larger," ASME Journal of Engineering for Power, Vol. 85, pp. 249–298, 1963.

velocity is known to be a substantial contribution to the total energy, but it is ignored in the *Exhaust Loss Method*, or rather the effect of it is "lumped" together in the *Exhaust Loss Correction*, which is used as an enthalpy adjustment. This can also be viewed as an efficiency correction.

The average (or *bulk*) steam velocity is used along with an *Exhaust Loss Curve* to arrive at an adjustment (or correction). Although the basis is vague and lacking in rigor, this method has proven to be quite successful-so much so that it is used throughout the industry worldwide and so it will be used here. The exhaust loss curve for the LP turbine in this example is shown in the following figure, along with the equations to be used in the calculation, including an adjustment for moisture. Curves similar to this are used throughout the industry.

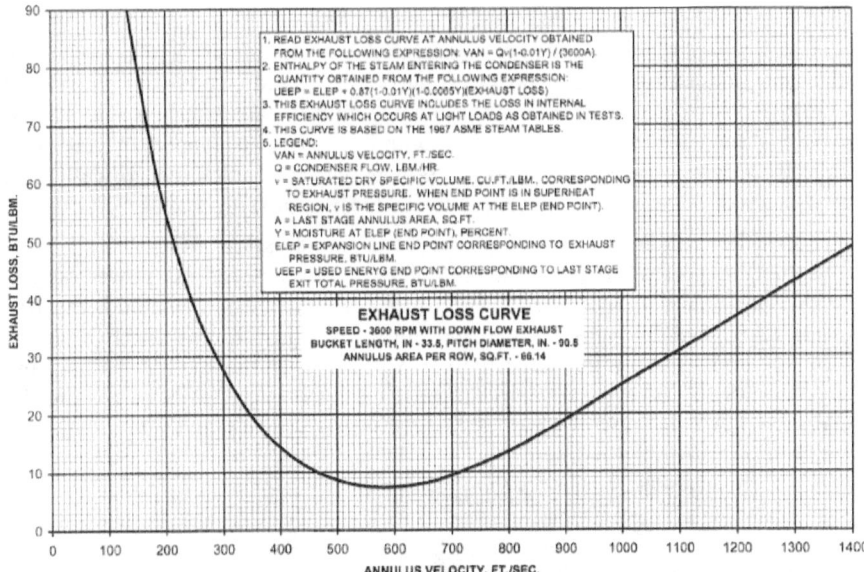

The exhaust loss can be used to calculate the ELEP from the UEEP, adding an additional equation and eliminating an additional unknown. There is still one more unknown than equation. Something else must be assumed. There are two approaches used: 1) match the slope of the expansion line segment from the hot reheat to the crossover and from the LPT inlet to the ELEP or 2) match the isentropic efficiency for these two expansions. The former yields two different efficiencies and the latter yields a single value. The spreadsheet has been set up to facilitate either approach. The calculations are shown in this next figure:

	A	B	C	D	E	F	G	H	I	J	K
1	design					Area [sq.ft.]	66.14				
2	measured						Vg	342.2952			
3	calculated						x	0.938523			
4	iterative						Van	1264.862			
5	solve						loss	42.17177	35.22332		
6	description	main steam	cold reheat	hot reheat	IPT exit	LPT inlet	exp.lin.end	used energ	IPT isentr.	LP admis.	LPT isentr.
7	point	1	2	3	4	6	ELEP	UEEP	4s	5	7s
8	flow		1778199			1812039	1874969			62930	
9	pressure	1820.9	618.6	551.7	97.32	97.32	0.9728	0.9728	97.32	97.32	0.97
10	temperatur	997.1	710.5	1006.2	590.5	589.5			531.3	558.2	100.8
11	enthalpy	1478.8	1356.0	1522.8	1324.9	1324.4	1006.1	1041.3	1295.3	1308.8	979.9
12	entropy	1.5727	1.5888	1.7282	1.7573	1.7567	1.8034	1.8663	1.7282	1.7417	1.7567
13	kW		64021.9			105082.1		155532.3			
14					slope, dh/ds	-6815	-6815	shaft	324636.4		
15					for method 1 adjust until	this is zero	0	PF	0.999224		
16						efficiency	86.98%	92.40%	loss	2318		
17					for method 2 adjust until	this is zero	5.42%	net calc	322318		
18								net meas	322318		
19					adjust until this is zero for either method				0		

The cells are color-coded as indicated in the upper left corner. The enthalpies and entropies are provided through function calls and the steam property Add-In. The shaft power for each section is calculated on row 13 and the sum shown in cell H14. The generator losses are calculated in cell H16, using the previous curves, and the net power output is calculated in cell H17. The measured value is in cell H18 and the difference in H19.

For either method, the Solver[33] is used to adjust the UEEP (cell H11, highlighted in yellow) to match the calculated and measured net power output. The two are matched when cell H19 equals zero. In the first method, the IPT exit enthalpy (cell E11, highlighted in yellow) is adjusted until the two slopes (cells E14 and F14) are the same (cell F15 is zero). In the second method, the IPT exit enthalpy is adjusted until the efficiencies (cells E16 and F16) are the same (cell F17 is zero). This second objective is achieved by adding a constraint, as illustrated below:

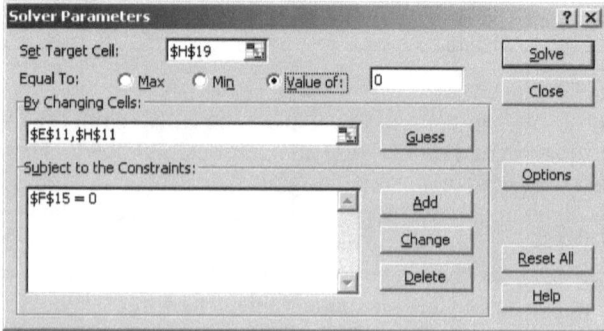

---

[33] The Solver is an Add-In function that comes with Excel®. In recent versions this can be found on the Data tab at the far right. On earlier versions it's on the Tools drop-down menu. You must enable this at some point, which is done on the File/Add-Ins tab or the Tools drop-down.

Chapter 14. Regenerative Rankine Cycle

The basic vapor power cycle is called the *Rankine Cycle*.[34] The Rankine Cycle takes advantage of the fact that the density of a liquid (most often water) is much greater than the density of the corresponding vapor. As specific volume is the inverse of density, this is the same as the specific volume of the liquid being much less than that of the vapor.

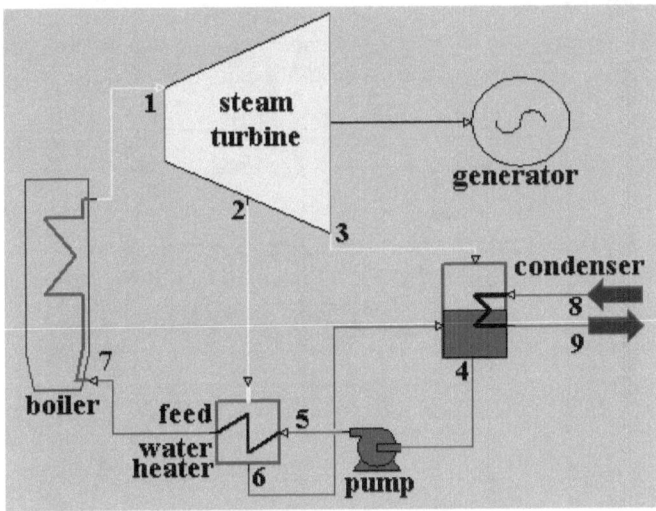

The working fluid is in the vapor state at points 1-3 (cyan streams) and in the liquid state at points 4-9 (blue streams). From Equation 1.3 we note that power is proportional to mass flow rate times specific volume times pressure differential. The difference in pressure across the turbine (i.e., between points 1 and 3) is equal to the difference in pressure across the pump (4 to 5) plus piping and miscellaneous losses.

If the volumetric flow rate of the vapor were not orders of magnitude greater than the liquid, it would require as much power to drive the pump as would be produced by the turbine and this cycle would be useless. In this case the power required by the pump is less than 2% of that produced by the turbine.

For the cycle as pictured here with typical operating conditions for steam, the volumetric flow rate at point 3 is 19,000 times the volumetric flow rate at point 4. Steam expands so greatly that the volumetric flow rate at point 3 is 600 times that at point 1 and 60 times that at point 2. This is one of several reasons that steam is by far the best working fluid for power production systems.

---

[34] After William John Macquorn Rankine (1820–1872), a Scottish mechanical engineer who also contributed to civil engineering, physics, and mathematics.

The regenerative aspect of this cycle is the extraction from the middle of the steam turbine to the feed water heater. The larger the temperature difference is during the heating process (as the liquid condensate is brought up to boiling and then superheating before entering the turbine) the more entropy will be generated and the greater the irreversibility (lost work) will be. It is more efficient to use some of the steam (i.e., the extraction from the middle of the turbine) to heat the feed water.

There is another way of looking at the impact of regeneration through the use of extraction steam that can be seen in the temperature-entropy diagram for this cycle:

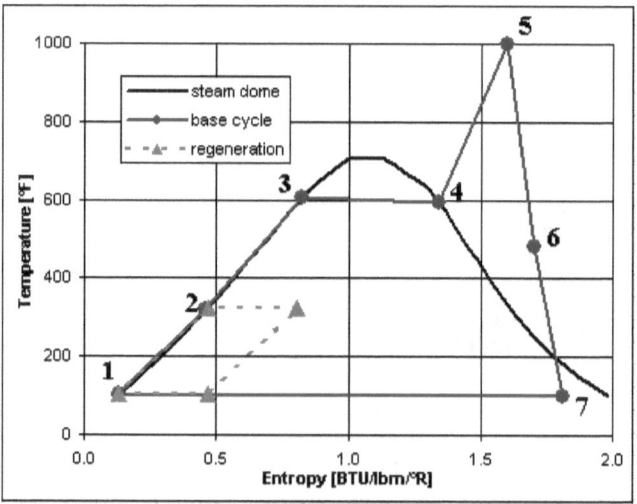

In this figure the blue line (with circular points) represents the base cycle without regeneration and the red line (with triangular points) represents the regeneration accomplished by the extraction. Regeneration is in effect cutting a triangle off of the bottom left corner of the base cycle and flipping it up into the middle. The end result is closer to a rectangle (i.e., the Carnot Cycle and best achievable efficiency) than before.

The process following points 1-5 is at nearly constant pressure. The shape of this process line is determined by the thermodynamic properties of steam, which are like any other fluid under corresponding conditions. The process following points 4-6 is determined by the efficiency of the turbine. When using a real working fluid, such as water or a refrigerant, you can't arbitrarily control the shape of this cycle.

It might seem from the preceding argument and the shape of the Rankine Cycle in the figure above that the easiest way to make this cycle more rectangular would be to lop off the upper right corner (point 5). There are two reasons for not doing this: 1) that would reduce the maximum temperature and the overall efficiency, and 2) there would be too much moisture in the steam at the exit of the turbine.

The black curve is the *steam dome* or the locus of saturated liquid and vapor points. When the blue line on the right is below the black curve, the steam is wet. High velocity water droplets will rapidly erode turbine blades and can cause catastrophic vibrations. Superheating the steam (i.e., raising it to 1000°F in this case between points 4 and 5) is necessary to provide the longest possible expansion line (points 5-6-7) before the steam becomes too wet.

Chapter 15. Gas Turbine Heat Balance

Performing a heat balance around a gas turbine is far more complicated than it may seem. Not only are there multiple streams entering and leaving, one must also consider the chemistry. The most common mistake made in analyzing such systems is mixing enthalpies from different sources with different reference points.

The properties of moist air are defined by the American Society of Heating, Refrigeration, and Air Conditioning Engineers (ASHRAE) in their Handbook of Fundamentals, which is published periodically. This is the de-facto worldwide standard. The reference point for these properties is as follows: 1) The enthalpy of dry air is zero at 0°F; and 2) The enthalpy of water vapor is zero at the triple point (i.e., 32.018°F). This difference in reference points for the dry air and water vapor is not a problem as long as you are dealing with moist air at near atmospheric conditions and use this same formulation throughout. The problem comes when you mix it with something else.

Properties of the fuel (whether gas or liquid) are not based on either 0°F or 32.018°F. There are many sources for fuel properties and it is important to recognize that the sensible and chemical energies may be at different reference conditions. The *CRC Handbook of Chemistry and Physics* is one common reference. Primary industrial references include the American Society of Mechanical Engineers (ASME) Power Test Code (PTC) 22 (Gas Turbines) and 4.4 (Gas Turbine Heat Recovery Steam Generators).[35]

There are three primary sources for gas (not fuel) properties: 1) NASA Glenn[36], 2) JANAF Gas Tables[37], and 3) the VDI Wärmeatlas.[38] All three are excellent. They also have different reference points, including 25°C/77°F or 0°C/32°F. To further complicate matters, there are on-line sources that may use yet different references.

A gas turbine may also have water or steam injection, which is done to control NOx emissions and increase performance. It is essential to recognize that

---

[35] Note that these two are not exactly the same in spite of statements to that effect. Note also that both of these claim to derive properties from the NASA Glenn Report as well as (Gas Producers Association Report) GPA-2145. While this is loosely true, it is not literally accurate. If you compare the numbers listed in the ASME documents to the NASA and GPA documents, they are not the same. This is due to conversions and adjustments that have been made by ASME.
[36] McBride, B. J., Zehe, M. J., Gordon, S., "NASA Glenn Coefficients for Calculating Thermodynamic Properties of Individual Species," NASA Report No. 211556, 2002.
[37] Chase M. W., Davies, C. A., Downey, J. R., Frurip, D. J., McDonald, R. A., and Syverud, A. N., "JANAF [Joint Army, Navy, Air Force] Thermochemical Tables, Third Edition,", Journal of Physical Chemistry Reference Data, Vol. 14, Supplement 1, 1985.
[38] Mewes, D., VDI-Wärmeatlas [Heat Atlas], 8th edition. VDI [Association of German Engineers] Society for Chemical Engineering, Springer-Verlag, 1997.

the steam (or water) properties do not have the same reference point as the water vapor enthalpies provided by NASA Glenn, JANAF, or VDI. If you subtract the exit enthalpy for water vapor in the exhaust from the steam enthalpy of the injection, you will introduce a significant error.

To further complicate matters, the heat of combustion is also dependent on reference temperature, as noted in the following figure, excerpted from a common engineering handbook:

Formula	State	Heat of combustion $-\Delta Hc°$, at 25°C. and constant pressure, to form					
		H_2O (liq.) and CO_2 (gas)			H_2O (gas) and CO_2 (gas)		
		Kcal./mole	Cal./g.	B.t.u./lb.	Kcal./mole	Cal./g.	B.t.u./lb.
H_2	gas	68.3174	33,887.6	60,957.7	57.7979	28,669.6	51,571.4
C	solid, graph.	94.0518	7,831.1	14,086.8			
CO	gas	67.6361	2,414.7	4,343.6			
Paraffins							
CH_4	gas	212.798	13,265.1	23,861	191.759	11,953.6	21,502
C_2H_6	gas	372.820	12,399.2	22,304	341.261	11,349.6	20,416
C_3H_8	gas	530.605	12,033.5	21,646	488.527	11,079.2	19,929
C_3H_8	liq.*	526.782	11,946.8	21,490	484.704	10,992.5	19,774
C_4H_{10}	gas	687.982	11,837.3	21,293	635.384	10,932.3	19,665
C_4H_{10}	liq.*	682.844	11,748.9	21,134	630.246	10,843.9	19,506
C_4H_{10}	gas	686.342	11,809.1	21,242	633.744	10,904.1	19,614
C_4H_{10}	liq.*	681.625	11,727.9	21,096	629.027	10,822.9	19,468
C_5H_{12}	gas	845.16	11,714.6	21,072	782.04	10,839.7	19,499
C_5H_{12}	liq.	838.80	11,626.4	20,914	775.68	10,751.5	19,340
C_5H_{12}	gas	843.24	11,688.0	21,025	780.12	10,813.1	19,451

You *must* use the same reference for all of the properties *and* the heating value; otherwise, you will introduce errors into the calculation. The following figure illustrates this problem:

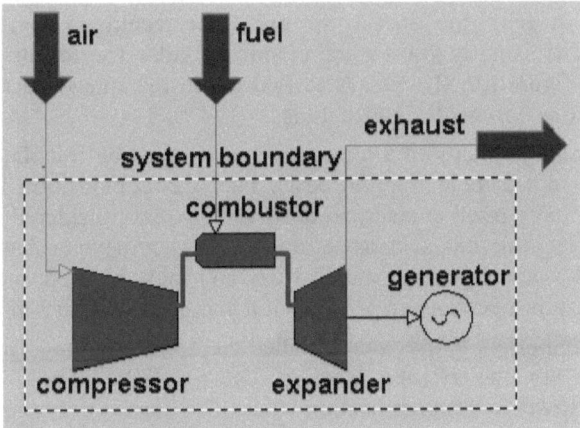

The following simplifications will be made to facilitate this example:

1) The fuel is pure methane.
2) There is no water or steam injection.
3) There is no bleed off of the compressor.
4) There is no gear loss (i.e., a dual-shaft engine).

5) The impact of fuel pressure on sensible heat is negligible.
6) The trace gases found in air (Hydrogen, , Neon, Krypton, Xenon Methane, and Sulfur Dioxide) can be lumped together with Helium, Argon, or Carbon Dioxide.
7) Complete combustion without NOx formation.
8) The heat loss can be characterized by an assumed percentage of the total heat input.

This problem is solved by first calculating the amount of oxygen required for combustion of the fuel. This has been hardwired into the Excel® spreadsheet for methane, but can be generalized using a table, as in the more complete spreadsheet, which is also included in the on-line archive. The fuel composition can also be generalized using tables (also in the more complete example).

The enthalpy is then calculated for each of the streams entering and leaving. This is done by considering the dry air required for stoichiometric combustion, the excess dry air, the moisture (water vapor) brought in with the dry air, the fuel, and each of the products of combustion as separate *streams*. These *streams* are entirely mixed, but the chemical species are considered separate and properly accounted for, so that the end result is the same. All of the enthalpies-including the moist air at the inlet-are calculated at the same reference condition: 25°C/77°F.

By considering the dry air required for combustion separate from the excess air, this adds an equation and eliminates the necessity of an iterative solution, as is the case if the combustion and excess air are considered jointly. The water vapor in the exhaust is equal to that brought in with the dry combustion air plus that brought in with the excess air plus that resulting from hydrocarbon combustion. The same is true for the carbon dioxide. The results are shown in the following figure for SI units. A second tab in the spreadsheet provides the same calculations only with English units.

Note that the enthalpy of a mixture of gases must be calculated on a mass fraction basis, not a mole fraction basis. Fuel gas composition is most often provided on a mole fraction basis, so this must also be considered when solving this problem for a mixture, such as natural gas. Also remember that the humidity ratio (mass of water vapor per mass of dry air) must be used and not relative humidity, which is a percentage of saturation and not a mass-based property.

It is also important to use what is called the *Lower Heating Value* (LHV) of the fuel and not the *Higher Heating Value* (HHV). There are different definitions of lower and higher heating value. The correct definition depends on whether the water resulting from combustion is in the *vapor* or *liquid* state. The vapor state produces the LHV and the liquid state produces the HHV. Notice that the header of the preceding table says, "to form," and is divided into two sections. The one on the left says, "H_2O (lqd.) and CO_2 (gas)," and the one on the right says, "H_2O (gas) and CO_2 (gas)." Unless you're burning fuel in

Antarctica during a blizzard, the H₂O will be in the vapor phase; so HHV is a meaningless quantity-don't use if for anything.

Gas Turbine Heat Balance Calculations (Methane Fuel Only)

Inputs	Value	Units	Calculations	Value	Units	Exhaust Mass Fractions	Units
barometric pressure	101.0	kPa	humidity ratio	0.015321	kg/kg	Nitrogen	72.9043%	-
ambient temperature	35.0	°C	shaft power	169,591	kW	Carbon Dioxide	5.4881%	-
ambient relative humidity	43%	-	heat losses	3,854	kW	Water (vapor)	5.9342%	-
fuel flow (methane)	9.450	kg/s	**Reactants**	**Value**	**Units**	Oxygen	14.4273%	-
fuel temperature	185	°C	Carbon Dioxide	25.924	kg/s	Argon	1.2461%	-
exhaust temperature	600	°C	Water (vapor)	21.224	kg/s	Helium	0.0001%	-
generator output	167,700	kWe	stoichiometric O2	37.699	kg/s	total	100.0000%	-
generator efficiency	98.89%	-	dry air	162.902	kg/s	**Enthalpies**	**Value**	**Units**
heat loss	0.82%	-	moist air	165.398	kg/s	ambient dry air	10.05	kJ/kg
fuel lower heating value	50,000	kJ/kg	**Products**	**Value**	**Units**	ambient moisture	18.04	kJ/kg
Dry Air Constituents by mole		**Units**	Carbon Dioxide	26.003	kg/s	fuel (sensible)	432.21	kJ/kg
Nitrogen	78.0842%	-	Water (vapor)	21.224	kg/s	excess air at Tex	605.30	kJ/kg
Oxygen	20.9477%	-	Nitrogen	123.022	kg/s	exhaust moisture	1144.23	kJ/kg
Argon	0.9359%	-	Argon	2.103	kg/s	products	696.55	kJ/kg
Carbon Dioxide	0.0316%	-	Helium	0.000	kg/s	exhaust	640.02	kJ/kg
Helium	0.0006%	-	total	172.352	kg/s			
total	100.0000%	-	excess dry air	297.001	kg/s			
Molecular Weights	**Value**	**Units**	excess moist air	301.552	kg/s			
Hydrogen	2.016	1/mole	total air (moist)	466.950	kg/s			
Helium	4.003	1/mole	total exh. (moist)	476.400	kg/s			
Carbon	12.011	1/mole	**Exhaust Constituents**		**Units**			
Nitrogen	28.013	1/mole	Nitrogen	347.316	kg/s	**Legend**		
Oxygen	31.999	1/mole	Carbon Dioxide	26.145	kg/s	user inputs		
Argon	39.948	1/mole	Water (vapor)	28.270	kg/s	calculated values		
Methane	16.042	1/mole	Oxygen	68.732	kg/s			
Water	18.015	1/mole	Argon	5.937	kg/s			
Carbon Dioxide	44.010	1/mole	Helium	0.000	kg/s			
Air (dry)	28.965	1/mole	total	476.400	kg/s			

Most of the equations in this spreadsheet are fairly simple, for example, mass flows times ratios of molecular weights and heating value times fuel flow. Even the dry air required for stoichiometric combustion is straightforward. The complicated equation arises when solving the overall energy balance for the flow of excess dry air, which is provided below:

$$m_{excess\_dry} = \frac{\begin{pmatrix} m_{fuel}(LHV + h_{fuel, Tfuel} - h_{products, Texh}) - Q_{loss} - W_{shaft} \\ -\frac{(h_{air, Tbleed} - h_{air, Tin} + \omega (h_{H2O, Tbleed} - h_{H2O, Tin})) m_{bleed\_total}}{1 + \omega} \\ - m_{comb\_dry}(h_{products, Texh} - h_{air, Tin} + \omega (h_{H2O, Texh} - h_{H2O, Tin})) \\ - m_{inject}(h_{H2O, Texh} - h_{H2O, Tinj}) \end{pmatrix}}{(h_{air, Texh} - h_{air, Tin} + \omega (h_{H2O, Texh} - h_{H2O, Tin}))}$$

The derivation of this equation is complicated and errors are easily made, which is why I used Maple® (the symbolic engine inside MathCAD®) to solve it. The Maple® script file (gas_turbine_combustion.mws) is included in the on-line archive, if you're interested and have software that can read it. The variables are as follows:

symbol	description
$h_{air,Tbleed}$	enthalpy of bleed air at T_{bleed}
$h_{air,Texh}$	enthalpy of air at T_{exh}
$h_{air,Tin}$	enthalpy of air at inlet temperature
$h_{fuel,Tfuel}$	enthalpy (sensible heat) of fuel at T_{fuel}
$h_{H2O,Tbleed}$	enthalpy of water vapor at T_{bleed}
$h_{H2O,Texh}$	enthalpy of water vapor at T_{exh}
$h_{H2O,Tin}$	enthalpy of water vapor at T_{in}
$h_{H2O,Tinj}$	enthalpy of water/steam injection
$h_{products,Texh}$	enthalpy of products at T_{exh}
LHV	fuel lower heating value
m_{bleed}	compressor bleed mass flow rate
m_{comb_dry}	mass flow rate of dry air required for combustion
m_{excess_dry}	excess air flow (dry)
m_{fuel}	fuel mass flow rate
m_{inject}	water/steam injection flow rate
Q_{loss}	heat loss
T_{bleed}	compressor bleed temperature
T_{exh}	exhaust temperature
T_{fuel}	fuel temperature
T_{in}	air inlet temperature
T_{inj}	water/steam injection temperature
ω	humidity ratio of ambient air
W_{shaft}	shaft power = generator output/efficiency

Chapter 16. Simple Combined Cycle

The simplest combined cycle is illustrated in the following figure. This system contains a combustion gas turbine generator (CTG), a heat recovery steam generator (HRSG), a steam turbine generator (STG), a condenser, and a pump. The HRSG consists of an economizer, an evaporator, and a superheater. The economizer heats the feed water from the pump up to near the boiling point. The evaporator continues this heating process up to the point of saturated vapor. The superheater continues to heat the steam to lengthen the expansion line, as described earlier in the Rankine Cycle.

The HRSG is broken up into these three components for physical and mechanical reasons. The liquid in the economizer and vapor in the superheater are both single phase and a similar piping design is used for these heat exchangers. The evaporator is where the boiling takes place. This is done in vertical columns so that the steam bubbles can rise up and separate from the liquid.

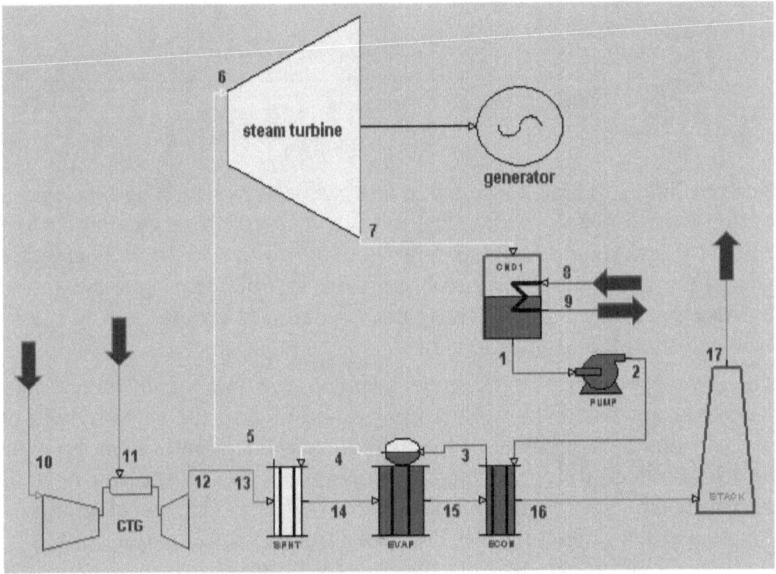

In a combined cycle plant, such as the one depicted here, the steam turbine functions the same as in a Rankine Cycle and the same machines are often used in both applications. The generator, condenser, and pumps, are also similar. The gas turbines are the same as are deployed without a HRSG. There are some similarities between a conventional boiler and the economizer-evaporator-superheater elements used in a HRSG, but these are not at all interchangeable.

The steam process line from points 1 through 7 is the basic Rankine Cycle without regeneration. In the case of a combined cycle, it is not advantageous to extract steam or use feed water heaters. The whole purpose of the HRSG is to extract heat from the gas turbine exhaust. Heat transfer in the HRSG between the gas turbine exhaust stream and the feed water-to-steam stream. It is informative to plot this process on a graph having heat transfer on the horizontal axis and temperature on the vertical, as illustrated below:

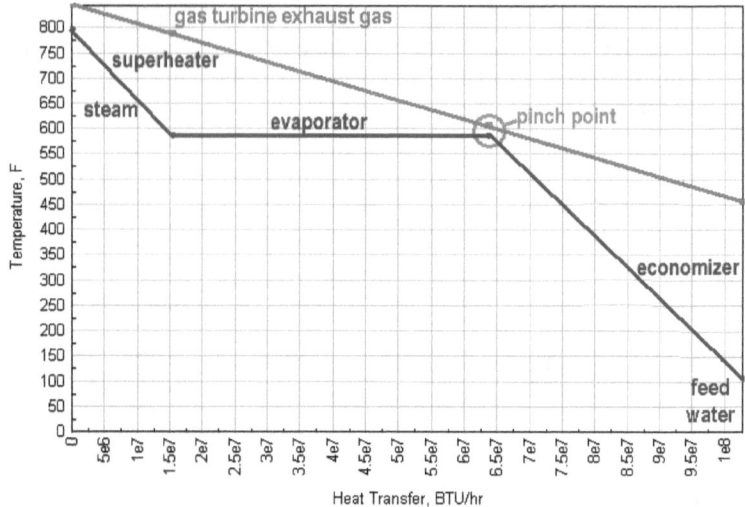

The red line is the gas turbine exhaust gas process line. The blue line is the feed water-to-steam process line. The exhaust gas process begins hot on the left at zero heat transfer and continues down (colder) to the right in a nearly straight line. This line would be exactly straight if the specific heat were constant. The steam process starts at the bottom right and continues upward and to the left as heat is added from the exhaust stream.

The left most segment of the blue steam process line identified as occurring in the economizer has a slope inversely proportional to the specific heat of the liquid. The right most segment of the blue process line identified as occurring in the superheater has a slope inversely proportional to the specific heat of the vapor. The center section of the blue process line is flat, because this is the evaporator and the temperature is constant throughout the boiling process. The specific heat of saturated steam is infinite, so the inverse is zero and so is the slope.

Because the red process line is nearly straight and the blue process line is a sloping stair step, there will always be at least one point where the two lines are very close. This is called a *pinch point*. The two lines can never touch, because heat is that form of energy that crosses a system boundary by virtue of a temperature difference. If there's no temperature difference, there is no heat transfer.

We recognize that the rate of heat transfer (dQ/dt) is proportional to temperature difference (ΔT) times surface area (A) times some factor of performance (i.e., a heat transfer coefficient, ζ). Solving for area yields:

$$A = \frac{\left(\dfrac{dQ}{dt}\right)}{\zeta \cdot \Delta T} \tag{16.1}$$

As the pinch becomes tighter and tighter (i.e., the temperature difference at the pinch point becomes smaller and smaller), the numerator in Equation 16.1 goes to zero, making the area required to transfer a desired quantity of heat larger and larger. Heat exchangers are basically sold by the pound (or kilogram), that is, their price is determined by materials plus fabrication. Bigger area means more money.

The pinch point in a HRSG is a thermodynamic bottleneck and an expensive one. There will always be some degree of pinch in a HRSG, because of the shape of the process lines, which is determined by the properties of exhaust gas and steam. The challenge is to manage the pinch and arrive at the most cost-effective design, which is a science in itself and beyond the scope of this book.

Chapter 17. Vapor-Compression Refrigeration Cycle

The vapor-compression refrigeration cycle is similar to the Rankine Cycle, only running backwards. In the Rankine Cycle a fluid such as water is boiled and superheated in order to expand it and extract power. Then this same fluid is then condensed and pumped back to the boiler. The heat rejected from the cycle through the condenser is considered waste.

In a refrigerator or air conditioner, the objective is to remove heat and put it somewhere else. Rather than extracting power from such a system, we must supply power to operate this cycle. Energy naturally travels from hotter objects to colder ones, which is the process we call heat transfer. This is analogous to water running down hill.

We can install a paddlewheel (or more sophisticated device) to extract power from water running down hill. Refrigeration is analogous to pumping water up hill and this requires power, rather than producing it. Reverse cycle air conditioners are sometimes called heat pumps. In truth, air conditioners are always heat pumps, whether they're pumping it out or pumping it in. The following figure is a schematic of this process:

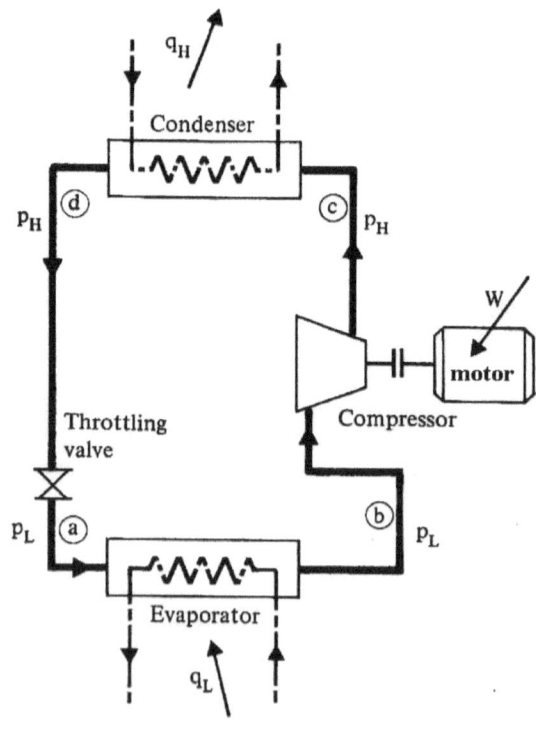

Energy (in the form of heat) is effectively *pulled* into the evaporator and *pushed* out through the condenser. The evaporator must be colder than what you want to refrigerate or heat won't flow out of your groceries and into the heat exchanger. Likewise, the condenser must be hotter than the surroundings where you want to dispose of the waste heat. The motor driving the compressor supplies the power required to accomplish this. The throttle valve is an abrupt pressure drop, where most of the pressure developed by the compressor is dissipated.

In an automobile air conditioner the compressor is driven by a belt. The evaporator sits behind the dashboard and gets cold. The condenser is in front of the radiator and gets hot. Fans blow air through the evaporator to cool the passenger compartment and through the condenser to dispose of the unwanted energy in the form of heat.

Conservation of energy around this system (i.e., the 1LoT) reveals that the heat rejected by the condenser, Q_H, is equal to the heat absorbed by the evaporator, Q_L, plus the power provided to the motor, W. This means that the refrigerator operating inside your kitchen is adding heat load to your house that your air conditioner must remove. The contents inside the refrigerator may be getting colder, but the kitchen is getting hotter, because there's a wire feeding power into the back of the refrigerator.

The ideal vapor-compression refrigeration cycle is shown on the following temperature-entropy chart:

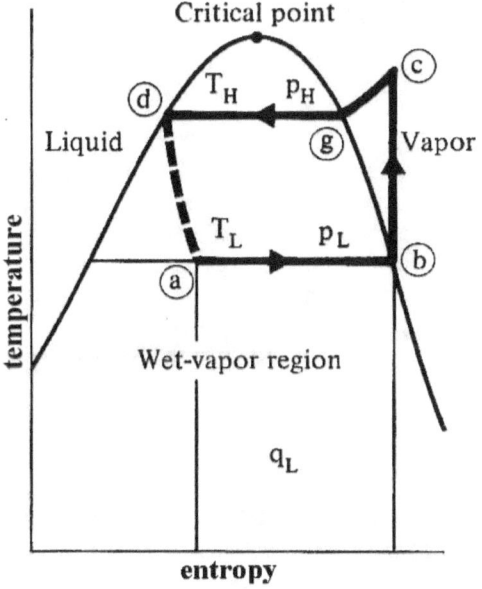

The process from point a to point b occurs in the evaporator, starting as a liquid with some vapor and continuing to the right until there is no more liquid. The process from point b to point c occurs in the compressor. This is all in the vapor phase. The process from point c to point d occurs in the condenser, beginning as a superheated vapor and continuing to the left until this is completely condensed. The process from point d to point a occurs in the expansion or throttling valve.

This is an ideal cycle in that the compression is shown as isentropic. This would require a compressor that was 100% efficient. Actual refrigeration and air conditioning compressors are rarely more than 70% efficient. The actual compression line (point b to point c) will not be vertical. There is also the efficiency of the motor to consider. Typical small single phase motors on the order of 1 kW (or 1 hp) are about 50% efficient.

This last step, expansion, is where the temperature plunges rapidly. While most fluids will exhibit a drop in temperature when expanded under some conditions, those conditions may not correspond to the ones you want for a refrigeration system. Some fluids exhibit a considerably larger temperature drop than others. There are many possible refrigerants. The ones most commonly used have been selected to provide the optimum performance at the desired conditions.

Expansion in the throttling valve occurs very rapidly. Ideally, this would be so rapid and occur over such a small area that there would be no opportunity for heat transfer. An expansion valve is an open system and produces no work; therefore, no heat transfer would mean no change in enthalpy. The rate of temperature with respect to pressure at constant enthalpy is called the Joule-Thompson Coefficient:[39]

$$\mu_{JT} = \frac{\partial T}{\partial P}\bigg)_{h=const.} \qquad (17.1)$$

We will analyze this refrigeration cycle in an Excel® spreadsheet that is included in the on-line archive and contains the complete thermodynamic properties for the 38 most common refrigerants. The spreadsheet is arranged so that you can select the refrigerant, the high and low temperatures, the cooling load, the compressor efficiency, and the motor efficiency. The properties are in tabular form, each on a separate tab. Excerpts from the property tabs are listed in Appendix C. Details of the cycle are shown in the following figure:

---

[39] After James Prescott Joule (1818-1889), an English physicist and brewer and William Thomson, 1st Baron Kelvin (1824–1907), a Scotch-Irish mathematical physicist and engineer.

	A	B	C	D	E	F
1	**Vapor-Compression Refrigeration Cycle**					
2	User Inputs		R134a	Refrigerant Number		
3	EvaporatorTemp.		-18	°C	0	
4	CondenserTemp.		74	°C	165	
5	Cooling Load		3.5	kW	1	
6	compressor eff.		70%	-		
7	motor efficiency		50%	-		
8	Calculations	T	P	h	s	x
9	point	°C	kPa	kJ/kg	kJ/kg/°K	quality
10	a	-18	146	163.2	0.6438	64%
11	b	-18	146	239.8	0.9438	100%
12	c	141	2304	322.7	1.0070	***
13	g	74	2304	280.6	0.8959	100%
14	d	74	2304	163.2	0.5572	0%
15	a	-18	146	163.2	0.6438	64%
16	b	-18	146	239.8	0.9438	100%
17	c'	101	2304	297.8	0.9438	***
18	flow	0.0459	kg/s			
19	compressor	3.81	kW			
20	electrical	7.62	kW			
21	coef.of perf.	0.462	-			

The blue cells are user inputs. You can select any one of the 38 refrigerants listed in Appendix C, change the high and low temperatures, or efficiencies and the calculations will automatically update along with the figures. This next figure is the T-s diagram for this cycle:

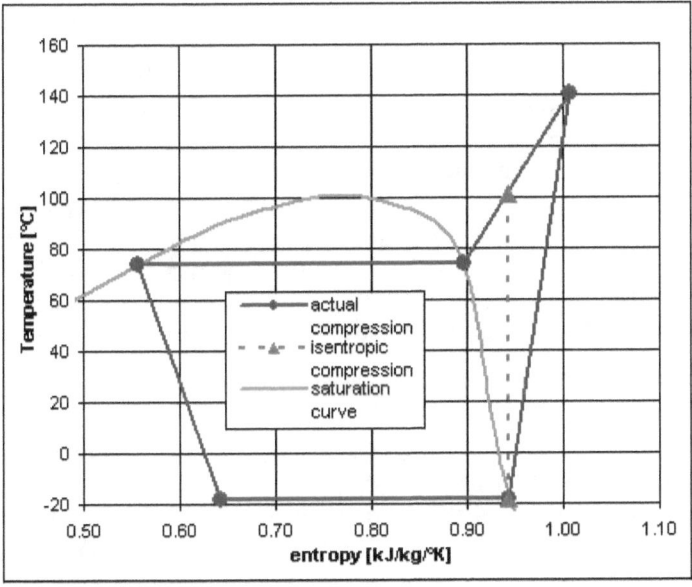

The green dashed line is the ideal compression and the solid blue line to the right of it is the actual compression, accounting for the efficiency of the compressor. Notice the two horizontal lines from points a to b and g to d. These are isothermal processes. Refrigeration cycles are often displayed on a pressure-enthalpy diagram, as in this next figure:

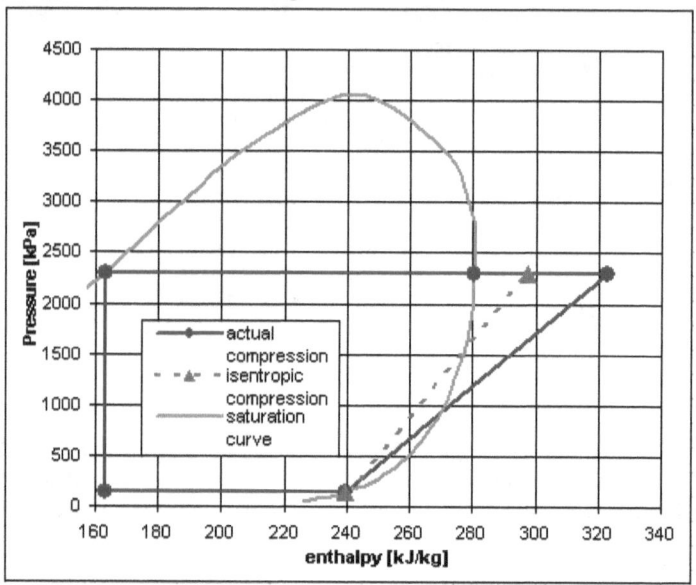

The vertical blue segment on the left is expansion through the throttle valve (i.e., from points d to a), which is presumed to be so rapid and occur in such a small device that there is no significant opportunity for heat transfer; thus, this process is isenthalpic (i.e., constant enthalpy). In this spreadsheet there is one tab for this cycle in SI units and another for English units.

The selected refrigerant for this example is R134a, which is commonly used in automotive air conditioners. The temperatures and heat loads selected are also common for a automotive applications. The coefficient of performance is calculated in cell B21 and is equal to 0.462 for this example. The coefficient of performance is defined as the heat removed divided by the power required, or:

$$cop = \frac{Q_L}{\dot{W}} \qquad (17.2)$$

The coefficient of performance is sometimes displayed on refrigerators and air conditioners. A significantly larger value than this is may be reported. In such cases, what they're not telling you is that they are showing is BTU/watt, which is 3.412 times the unitless value obtained from Equation 17.2.

Chapter 18. Otto Cycle

The ideal representation of a spark-ignition internal combustion engine is the Otto Cycle.[40] This is the cycle that gasoline engines operate on. Ideally it consists of 1) an isentropic (constant entropy) compression, 2) an isochoric (constant volume) heat addition, 3) an isentropic expansion, and 4) an isochoric heat rejection. In an actual engine the compression and expansion will achieve something less than 100% and will result in some increase in entropy.

A spreadsheet (cycles.xls) has been included in the on-line archive that makes the necessary calculations assuming ideal gas properties for air. The user inputs and resulting process lines are shown in the following figure:

Otto Cycle					
User Inputs					
compression ratio	9.5				
compression eff.	90%				
expansion efficiency	90%				
combustion temp.	1500	°C			
k=Cp/Cv	1.4	-			
Cp	1.1	kJ/kg/°C			
R	0.314	kJ/kg/°C			
point	P	T	v	h	s
units	kPa	°C	m³/kg	kJ/kg	kJ/kg/°C
1	101	25	0.925	0	0.000
1.1	138	55	0.746	33	0.008
1.2	183	85	0.614	66	0.015
1.3	238	115	0.513	99	0.021
1.4	302	145	0.434	132	0.027
1.5	378	175	0.372	164	0.033
1.6	466	204	0.322	197	0.038
1.7	568	234	0.281	230	0.043
1.8	684	264	0.247	263	0.048
1.9	815	294	0.219	296	0.052
2	963	324	0.195	329	0.057
2.1	1152	442	0.195	458	0.198
2.2	1342	559	0.195	588	0.317
2.3	1531	677	0.195	717	0.421
2.4	1721	794	0.195	846	0.513
2.5	1910	912	0.195	976	0.595
2.6	2100	1030	0.195	1105	0.669

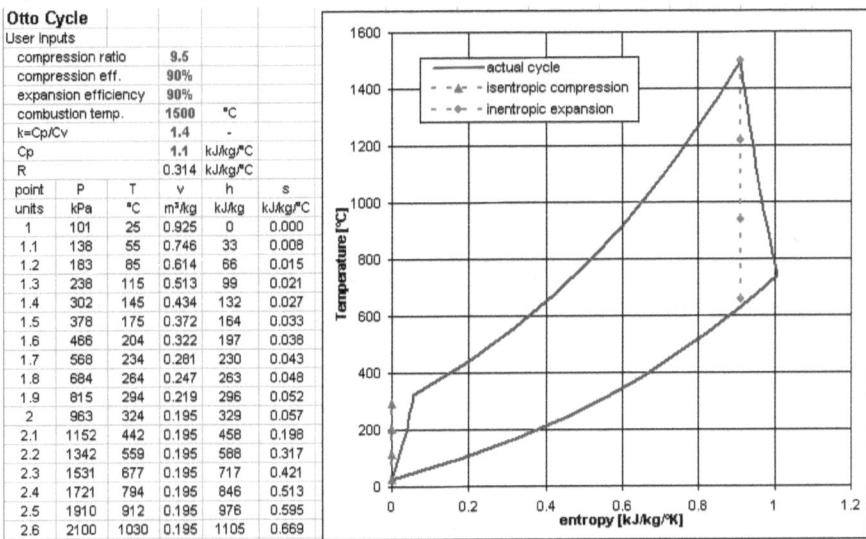

Typical inputs have been provided, including a compression ratio of 9.5:1 and a combustion temperature of 1500°C. Compression and expansion efficiencies of 90% are highly unlikely. These highly optimistic values have been selected to avoid grossly distorting the shape of the cycle.

---

[40] After Rudolf Christian Karl Diesel (1858-1913), a German thermal engineer who invented the compression-ignition internal-combustion engine.

These same process lines on a p-v diagram are shown in this next figure, where the constant volume heat addition and rejection become vertical lines:

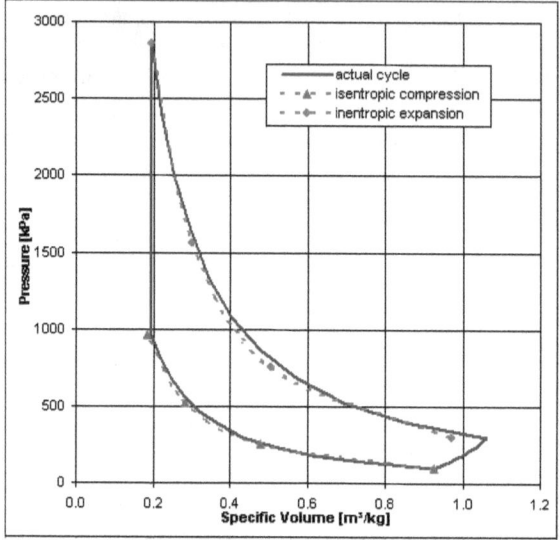

Approximations used for the air properties are given in Appendix D.

Chapter 19. Diesel Cycle

The ideal representation of a compression-ignition internal combustion engine is the Diesel Cycle.[41] This is the cycle that diesel engines operate on. Ideally it consists of 1) an isentropic (constant entropy) compression, 2) an isobaric (constant pressure) heat addition, 3) an isentropic expansion, and 4) an isochoric heat rejection. Except for the heat addition, it is identical to the Otto cycle. In an actual engine the compression and expansion will achieve something less than 100% and will result in some increase in entropy.

A spreadsheet (cycles.xls) has been included in the on-line archive that makes the necessary calculations assuming ideal gas properties for air. The user inputs and resulting process lines are shown in the following figure:

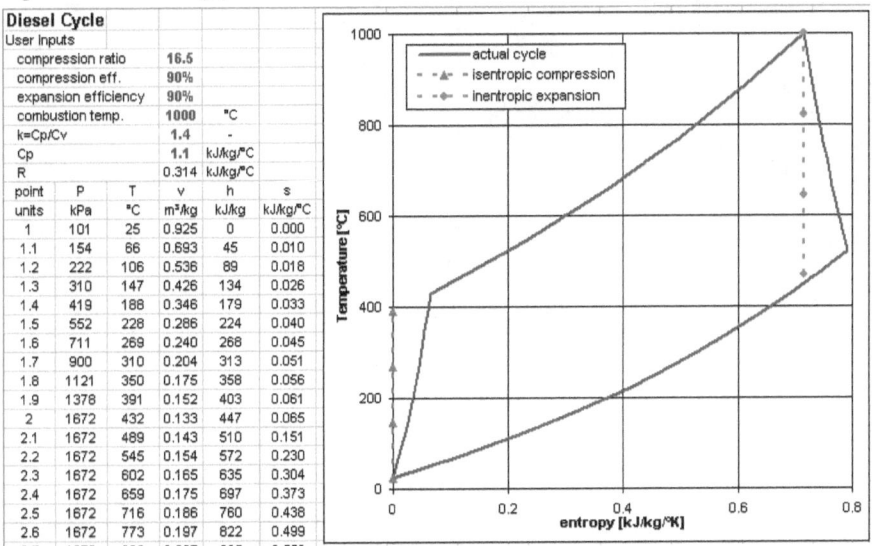

Typical inputs have been provided, including a compression ratio of 16.5:1 (significantly higher than a spark-ignition engine) and a combustion temperature of 1000°C. Compression and expansion efficiencies of 90% are highly unlikely. These highly optimistic values have been selected to avoid grossly distorting the shape of the cycle. Not only is the compression ratio of diesel engines much larger than for a gasoline engine, the maximum temperature is significantly lower, resulting in less NOx production.

---

[41] After Rudolf Christian Karl Diesel (1858-1913), a German thermal engineer who invented the compression-ignition internal-combustion engine.

These same process lines on a p-v diagram are shown in this next figure, where the constant pressure heat addition becomes a horizontal line and the constant volume heat rejection becomes a vertical line:

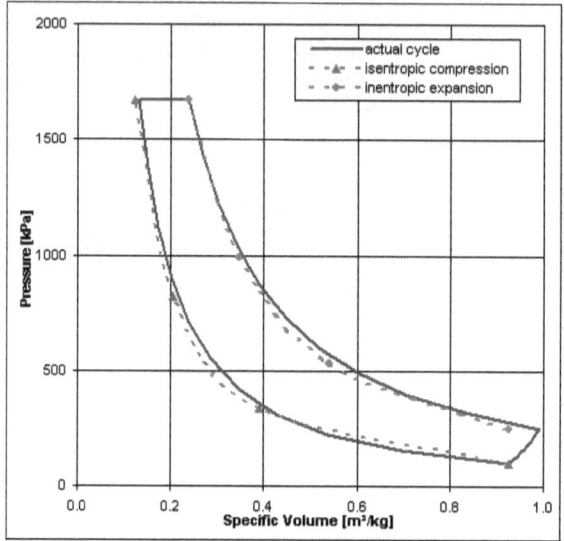

The Diesel Cycle requires an iterative solution to determine the state following the isentropic expansion. This solution is built into the spreadsheet.

Chapter 20. Brayton Cycle

The ideal representation of a constant pressure internal combustion engine is the Brayton Cycle.[42] This is the cycle that gas turbines operate on. Ideally it consists of 1) an isentropic (constant entropy) compression, 2) an isobaric (constant pressure) heat addition, 3) an isentropic expansion, and 4) an isobaric heat rejection. The Brayton Cycle is another variant, similar to the Otto and Diesel cycles. In an actual engine the compression and expansion will achieve something less than 100% and will result in some increase in entropy.

A spreadsheet (cycles.xls) has been included in the on-line archive that makes the necessary calculations assuming ideal gas properties for air. The user inputs and resulting process lines are shown in the following figure:

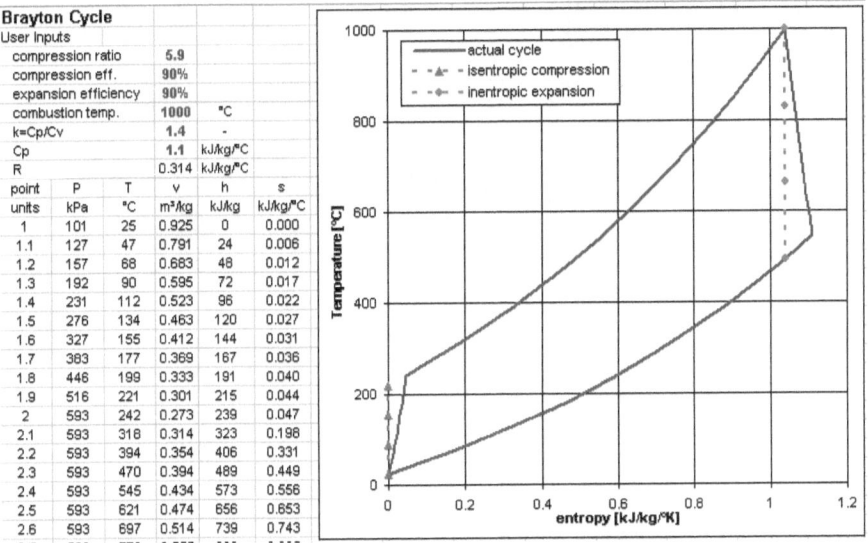

Brayton Cycle				
User Inputs				
compression ratio	5.9			
compression eff.	90%			
expansion efficiency	90%			
combustion temp.	1000	°C		
k=Cp/Cv	1.4	-		
Cp	1.1	kJ/kg/°C		
R	0.314	kJ/kg/°C		
point	P	T	v	h	s
units	kPa	°C	m³/kg	kJ/kg	kJ/kg/°C
1	101	25	0.925	0	0.000
1.1	127	47	0.791	24	0.006
1.2	157	68	0.683	48	0.012
1.3	192	90	0.595	72	0.017
1.4	231	112	0.523	96	0.022
1.5	276	134	0.463	120	0.027
1.6	327	155	0.412	144	0.031
1.7	383	177	0.369	167	0.036
1.8	446	199	0.333	191	0.040
1.9	516	221	0.301	215	0.044
2	593	242	0.273	239	0.047
2.1	593	318	0.314	323	0.198
2.2	593	394	0.354	406	0.331
2.3	593	470	0.394	489	0.449
2.4	593	545	0.434	573	0.556
2.5	593	621	0.474	656	0.653
2.6	593	697	0.514	739	0.743

---

[42] After George Brayton (1830-1892), an American mechanical engineer who developed the constant pressure engine that is the basis for the gas turbine.

These same process lines on a p-v diagram are shown in this next figure, where the constant pressure heat addition and rejection become horizontal lines:

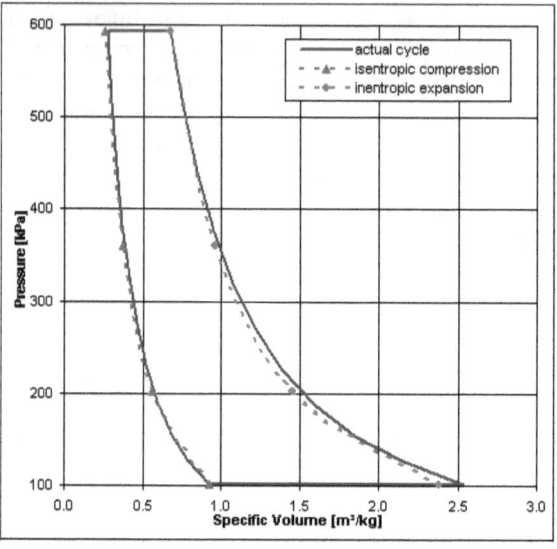

Chapter 21. Lenoir Cycle

The Lenoir cycle is similar to the Brayton cycle without the expansion and a constant volume, rather than constant pressure, heat addition.[43] It also has three instead of four points. This is the cycle that pulse jets operate on. Ideally it consists of 1) an isochoric (constant volume) heat addition, 2) an isentropic (constant entropy) expansion, and 3) an isobaric (constant pressure) heat rejection.

A spreadsheet (cycles.xls) has been included in the on-line archive that makes the necessary calculations assuming ideal gas properties for air. The user inputs and resulting process lines are shown in the following figure:

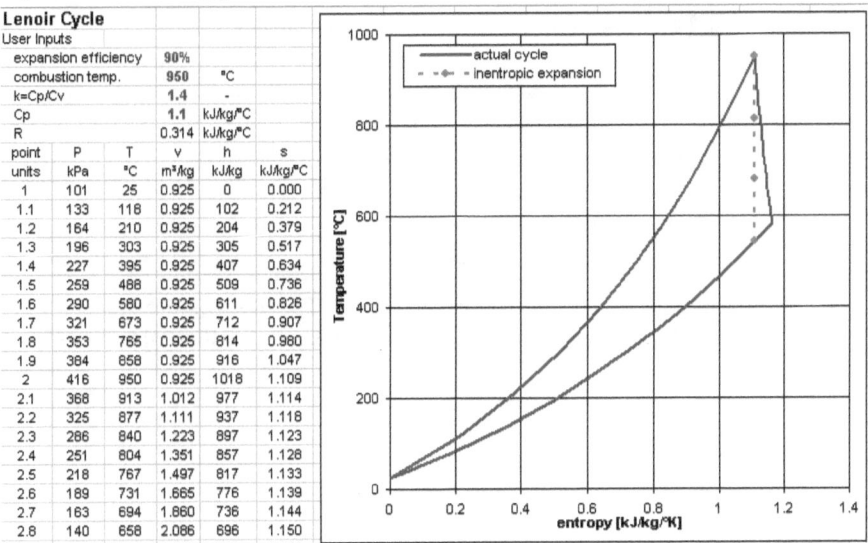

Lenoir Cycle					
User Inputs					
expansion efficiency	90%				
combustion temp.	950	°C			
k=Cp/Cv	1.4	-			
Cp	1.1	kJ/kg/°C			
R	0.314	kJ/kg/°C			
point	P	T	v	h	s
units	kPa	°C	m³/kg	kJ/kg	kJ/kg/°C
1	101	25	0.925	0	0.000
1.1	133	118	0.925	102	0.212
1.2	164	210	0.925	204	0.379
1.3	196	303	0.925	305	0.517
1.4	227	395	0.925	407	0.634
1.5	259	488	0.925	509	0.736
1.6	290	580	0.925	611	0.826
1.7	321	673	0.925	712	0.907
1.8	353	765	0.925	814	0.980
1.9	384	858	0.925	916	1.047
2	416	950	0.925	1018	1.109
2.1	368	913	1.012	977	1.114
2.2	325	877	1.111	937	1.118
2.3	286	840	1.223	897	1.123
2.4	251	804	1.351	857	1.128
2.5	218	767	1.497	817	1.133
2.6	189	731	1.665	776	1.139
2.7	163	694	1.860	736	1.144
2.8	140	658	2.086	696	1.150

---

[43] After George Brayton (1830-1892), an American mechanical engineer who developed the constant pressure engine that is the basis for the gas turbine.

These same process lines on a p-v diagram are shown in this next figure, where the constant pressure heat rejection becomes a horizontal line:

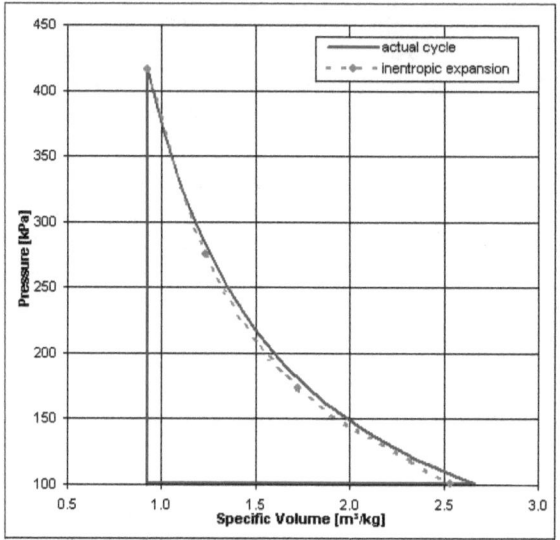

The Lenoir cycle is not particularly efficient. The T-s graph of the cycle is far from rectangular. Pulse jets were developed because of their simplicity, not their efficiency

Chapter 22. Stirling Cycle

The ideal representation of a constant temperature heat engine is the Stirling Cycle.[44] This cycle ideally it consists of 1) an isothermal (constant temperature) compression, 2) an isochoric (constant volume) heat addition, 3) an isothermal expansion, and 4) an isochoric heat rejection. The Stirling Cycle is not particularly useful. It's enticing, because it looks more like the Carnot Cycle than any of these others. The problem is building a practical engine.

A spreadsheet (cycles.xls) has been included in the on-line archive that makes the necessary calculations assuming ideal gas properties for air. The user inputs and resulting process lines are shown in the following figure:

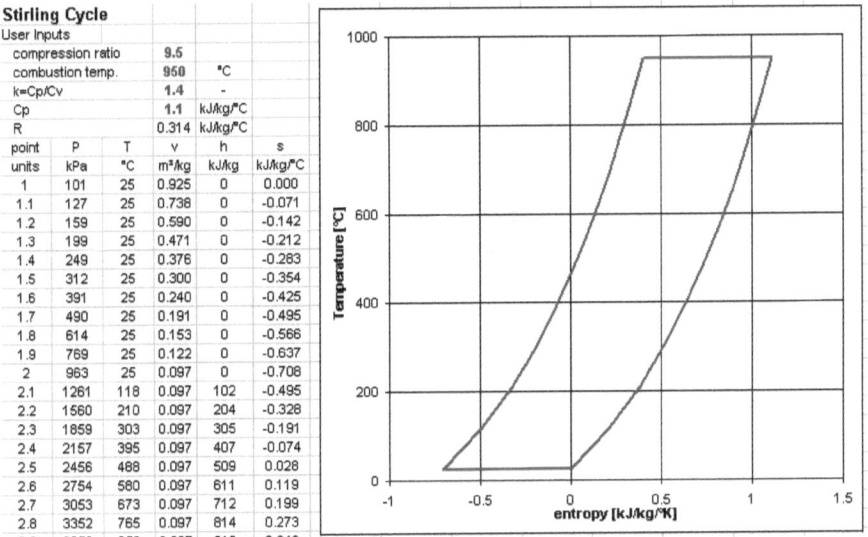

Stirling Cycle				
User Inputs				
compression ratio	9.5			
combustion temp.	950	°C		
k=Cp/Cv	1.4			
Cp	1.1	kJ/kg/°C		
R	0.314	kJ/kg/°C		
point	P	T	v	h	s
units	kPa	°C	m³/kg	kJ/kg	kJ/kg/°C
1	101	25	0.925	0	0.000
1.1	127	25	0.738	0	-0.071
1.2	159	25	0.590	0	-0.142
1.3	199	25	0.471	0	-0.212
1.4	249	25	0.376	0	-0.283
1.5	312	25	0.300	0	-0.354
1.6	391	25	0.240	0	-0.425
1.7	490	25	0.191	0	-0.495
1.8	614	25	0.153	0	-0.566
1.9	769	25	0.122	0	-0.637
2	963	25	0.097	0	-0.708
2.1	1261	118	0.097	102	-0.495
2.2	1560	210	0.097	204	-0.328
2.3	1859	303	0.097	305	-0.191
2.4	2157	395	0.097	407	-0.074
2.5	2458	488	0.097	509	0.028
2.6	2754	580	0.097	611	0.119
2.7	3053	673	0.097	712	0.199
2.8	3352	765	0.097	814	0.273

---

[44] After Robert Stirling (1790-1878), a Scottish clergyman and inventor. He was also a grandson of Michael Stirling, inventor of the threshing machine.

These same process lines on a p-v diagram are shown in this next figure, where the heat addition and rejection become vertical lines:

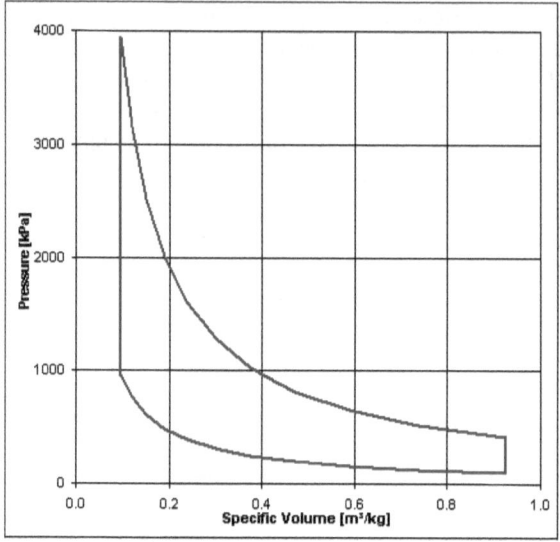

The problem with Stirling engines is that you can't get enough power out of them to be cost-effective.[45]

---

[45] When I was in graduate school, the department had a collection of antique engines. Among these was a Stirling engine about the size of a washing machine. One day when no one was looking, I fired it up-literally-I lit a gas fire under it. I oiled it and let it run for a while. Then I wanted to see how much power it could produce. I slid my hand along the flywheel and easily brought it to a stop. The power output was less than pathetic. If I had tried to stop the steam engine of similar size sitting next to the Stirling engine, it would have ripped my arms off.

Chapter 23. Ericsson Cycle

The Ericsson Cycle is yet another variant on these constant pressure, temperature, and volume processes.[46] This cycle ideally it consists of 1) an isothermal (constant temperature) compression, 2) an isobaric (constant pressure) heat addition, 3) an isothermal expansion, and 4) an isobaric heat rejection. Ericsson engines have not received much practical attention for the same reason as the Stirling engine, namely, it's virtually impossible to achieve an isothermal compression or expansion quickly.

A spreadsheet (cycles.xls) has been included in the on-line archive that makes the necessary calculations assuming ideal gas properties for air. The user inputs and resulting process lines are shown in the following figure:

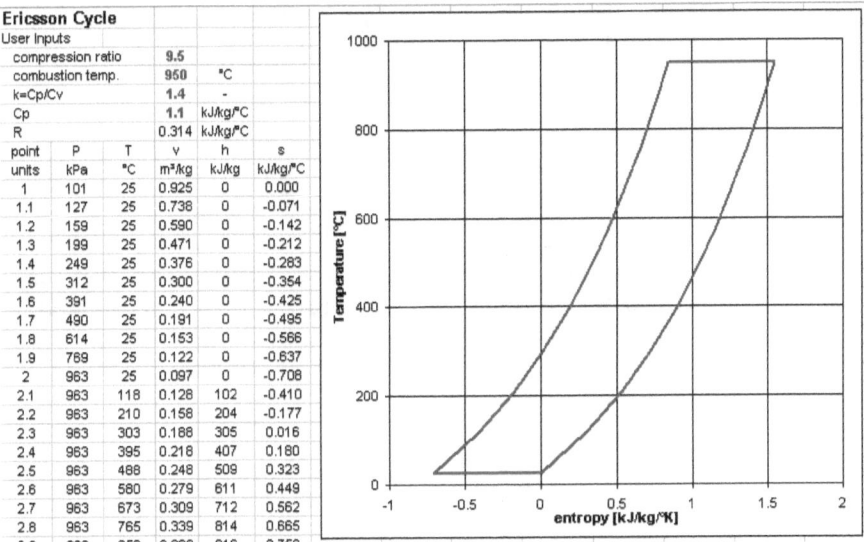

Ericsson Cycle					
User Inputs					
compression ratio	9.5				
combustion temp.	950	°C			
k=Cp/Cv	1.4	-			
Cp	1.1	kJ/kg/°C			
R	0.314	kJ/kg/°C			
point	P	T	v	h	s
units	kPa	°C	m³/kg	kJ/kg	kJ/kg/°C
1	101	25	0.925	0	0.000
1.1	127	25	0.738	0	-0.071
1.2	159	25	0.590	0	-0.142
1.3	199	25	0.471	0	-0.212
1.4	249	25	0.376	0	-0.283
1.5	312	25	0.300	0	-0.354
1.6	391	25	0.240	0	-0.425
1.7	490	25	0.191	0	-0.495
1.8	614	25	0.153	0	-0.566
1.9	769	25	0.122	0	-0.637
2	963	25	0.097	0	-0.708
2.1	963	118	0.128	102	-0.410
2.2	963	210	0.158	204	-0.177
2.3	963	303	0.188	305	0.016
2.4	963	395	0.218	407	0.180
2.5	963	488	0.248	509	0.323
2.6	963	580	0.279	611	0.449
2.7	963	673	0.309	712	0.562
2.8	963	765	0.339	814	0.665

---

[46] After Johan Ericsson (1803–1889), a Swedish-American inventor who also collaborated on the design of the steam locomotive.

These same process lines on a p-v diagram are shown in this next figure, where the heat addition and rejection become vertical lines:

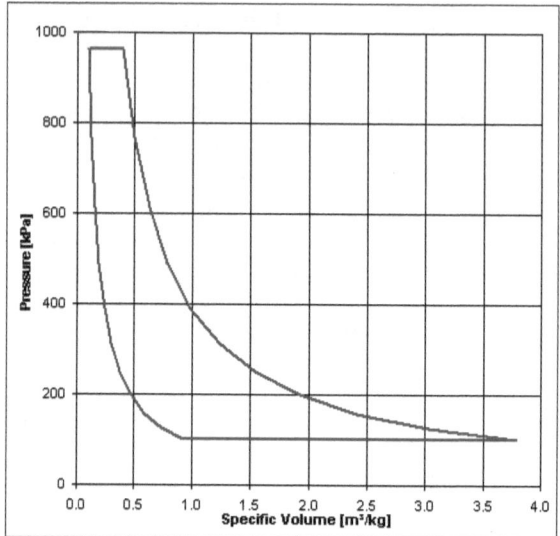

Notice that neither the Stirling nor the Ericsson cycle spreadsheets have inputs for compression or expansion efficiency (i.e., cells D4 and D5 in the Otto, Diesel, and Brayton spreadsheets). This is one of the problems with these cycles: when you try to account for actual, rather than ideal, performance of the machine components, the cycles degenerate into something else.

Chapter 24. Throttling Natural Gas

Throttling is a common thermodynamic process. We introduced a throttle and the Joule-Thompson coefficient in Chapter 17. Throttling can be a concern with natural gas, especially when it contains some moisture. This can result in condensation of the heavier components and even freezing of moisture. There is a spreadsheet in the on-line archive to analyze natural gas having 21 components, as listed below:

Natural Gas Mole Fractions and Properties											30	temperature [°F]
											600	pressure [psia]
name	Mole Fraction	formula	MW	LHV BTU/lb-mole	HHV BTU/lb-mole	Tc °R	Pc psia	Vc ft3/lbm	Zc	Partial-Pres. psia	Vapor Pres. psia	test cond-sed
1 Methane	91.900%	CH4	16.043	909.4	1010.0	343.0	667.8	0.09888	0.2878	551.400	N/A	N/A
2 Nitrogen	4.220%	N2	28.013	0.0	0.0	227.2	492.8	0.05151	0.2916	25.320	N/A	N/A
3 Carbon Dioxide	1.050%	CO2	44.010	0.0	0.0	548.2	1070.6	0.03423	0.2741	6.300	N/A	N/A
4 Ethane	1.810%	C2H6	30.070	1618.7	1769.6	549.8	707.8	0.07891	0.2846	10.860	337.494	no
5 n-Propane	0.610%	C3H8	44.097	2314.9	2516.1	665.7	616.3	0.07382	0.2808	3.660	66.771	no
6 Water	0.040%	H2O	18.015	0.0	0.0	1165.2	3208.1	0.04929	0.2278	0.240	0.080	YES
7 Hydrogen Sulfide	0.000%	H2S	34.082	586.8	637.1	672.4	1306.5	0.05167	0.3188	0.000	N/A	N/A
8 Hydrogen	0.000%	H2	2.016	273.8	324.2	59.8	188.1	0.51672	0.3053	0.000	N/A	N/A
9 Carbon Monoxide	0.000%	CO	28.010	320.5	320.5	239.3	507.5	0.05322	0.2946	0.000	N/A	N/A
10 Oxygen	0.000%	O2	31.999	0.0	0.0	278.6	736.9	0.03823	0.3015	0.000	N/A	N/A
11 i-Butane	0.120%	C4H10	58.123	3000.4	3251.9	734.6	529.1	0.07248	0.2827	0.720	21.944	no
12 n-Butane	0.160%	C4H10	58.123	3010.8	3262.3	765.3	554.0	0.07026	0.2755	0.960	14.376	no
13 i-Pentane	0.040%	C5H12	72.150	3699.0	4000.9	828.7	490.4	0.06787	0.2700	0.240	4.766	no
14 n-Pentane	0.020%	C5H12	72.150	3706.9	4008.9	845.4	490.1	0.06759	0.2634	0.120	3.391	no
15 n-Hexane	0.030%	C6H14	86.177	3856.6	4205.4	913.2	430.6	0.06875	0.2603	0.180	0.832	no
16 n-Heptane	0.000%	C7H16	100.204	4465.7	4864.3	972.3	396.8	0.06905	0.2631	0.000	0.204	no
17 n-Octane	0.000%	C8H18	114.231	5074.9	5523.4	1023.8	360.6	0.06905	0.2589	0.000	0.053	no
18 n-Nonane	0.000%	C9H20	128.258	5683.2	6181.6	1070.2	335.1	0.06630	0.2481	0.000	0.015	no
19 n-Decane	0.000%	C10H22	142.285	6294.7	6842.8	1111.9	304.6	0.06736	0.2447	0.000	0.004	no
20 Helium	0.000%	He	4.003	0.0	0.0	9.3	33.2	0.23115	0.3078	0.000	N/A	N/A
21 Argon	0.000%	Ar	39.948	0.0	0.0	271.5	706.9	0.02989	0.2897	0.000	N/A	N/A
Composite	100.000%		17.440	15,661	17,343	347.5	649.2	0.09486	0.2876	600.000		

The blue cells are user inputs and the violet cells are calculated values. The mixture is assumed to obey Dalton's Law of Partial Pressures.[47] The mixture is also assumed to follow the mixing rule and p-v-T behavior described by Soave's modification of the Redlich-Kwong EOS presented in Chapter 11.

Of particular interest is the possible condensation of any constituents. When the gas is throttled, the temperature will drop, due to the Joule-Thompson effect. Condensation could occur if the partial pressure of any of the constituents is greater than or equal to the vapor pressure at the same temperature. The result of this comparison for each component is shown in the far right column as, "YES," in case of water.

The spreadsheet can be used to calculate various throttling states, including isenthalpic (probable) and isentropic (highly unlikely). The spreadsheet can also be used to calculate a variety of state variables for the mixture, including

---

[47] After John Dalton (1766–1844), an English chemist, physicist, and meteorologist. This law (or rule) states that the constituents of an ideal mixture of gases each exert s fraction of the total pressure in direct proportion to their relative molar abundance.

density, critical properties of the mixture, and heating values. The following example shows the results of an isenthalpic expansion:

Throttling Calculations

State 1	State 2	property
80	45	temperature [°F]
1215	615	pressure [psia]
540	505	temperature [°R]
0.860	0.897	compressibility
0.235	0.453	specific volume [ft^3/lbm]
4.25	2.21	density [lbm/ft^3]
0.519	0.512	ideal gas specific heat [BTU/lbm/°F]
6.2	-11.8	ideal gas enthalpy [BTU/lbm]
-0.51	-0.467	ideal gas entropy [BTU/lbm/°R]
-35.3	-35.3	real enthalpy [BTU/lbm]
-0.46	-0.390	real entropy [BTU/lbm/°R]
	0.0	change in enthalpy [BTU/lbm]
	0.071	change in entropy [BTU/lbm/°R]

The spreadsheet includes the following 122 useful functions:

Mixture Properties
- CompositeCriticalCompressibility()
- CompositeCriticalPressure()
- CompositeCriticalTemperature()
- CompositeCriticalVolume()
- CompositeHigherHeatingValue()
- CompositeLowerHeatingValue()
- CompositeMolecularWeight()
- CompositePitzerAcentricFactor()
- CompositeRedlichKwongA()
- CompositeRedlichKwongB()
- CompositeRedlichKwongEm()
- CompositionName()
- SymbolicName()

Ideal Gas Properties
- IdealGasEnthalpy()
- IdealGasEntropy()
- IdealGasSpecificHeat()

Critical Properties
- CriticalCompressibility()
- CriticalPressure()
- CriticalTemperature()
- CriticalVolume()
- MolecularWeight()
- PitzerAcentricFactor()

Component Vapor Pressures
- VaporPressureButane()
- VaporPressureDecane()
- VaporPressureEthane()
- VaporPressureHeptane()
- VaporPressureHexane()
- VaporPressureIsoButane()
- VaporPressureIsoPentane()
- VaporPressureNonane()
- VaporPressureOctane()
- VaporPressurePentane()
- VaporPressurePropane()
- VaporPressureWater()

Component Heating Values
- LowerHeatingValue()
- HigherHeatingValue()

Real Gas Properties
- RealEnthalpy()
- RealEntropy()
- RedlichKwongA()
- RedlichKwongB()
- RedlichKwongCompressibility()
- RedlichKwongEm()
- RedlichKwongPressure()
- RedlichKwongVolume()

Component Specific Heats	Component Enthalpies	Component Entropies
SpecificHeatArgon()	EnthalpyArgon()	EntropyArgon()
SpecificHeatButane()	EnthalpyButane()	EntropyButane()
SpecificHeatCarbonDioxide()	EnthalpyCarbonDioxide()	EntropyCarbonDioxide()
SpecificHeatCarbonMonoxide()	EnthalpyCarbonMonoxide()	EntropyCarbonMonoxide()
SpecificHeatDecane()	EnthalpyDecane()	EntropyDecane()
SpecificHeatEthane()	EnthalpyEthane()	EntropyEthane()
SpecificHeatEthylene()	EnthalpyEthylene()	EntropyEthylene()
SpecificHeatHelium()	EnthalpyHelium()	EntropyHelium()
SpecificHeatHeptane()	EnthalpyHeptane()	EntropyHeptane()
SpecificHeatHexane()	EnthalpyHexane()	EntropyHexane()
SpecificHeatHydrogen()	EnthalpyHydrogen()	EntropyHydrogen()
SpecificHeatHydrogenSulfide()	EnthalpyHydrogenSulfide()	EntropyHydrogenSulfide()
SpecificHeatHydroxl()	EnthalpyHydroxl()	EntropyHydroxl()
SpecificHeatIsoButane()	EnthalpyIsoButane()	EntropyIsoButane()
SpecificHeatIsoPentane()	EnthalpyIsoPentane()	EntropyIsoPentane()
SpecificHeatMethane()	EnthalpyMethane()	EntropyMethane()
SpecificHeatNeon()	EnthalpyNeon()	EntropyNeon()
SpecificHeatNitrogen()	EnthalpyNitrogen()	EntropyNitrogen()
SpecificHeatNitrogenDioxide()	EnthalpyNitrogenDioxide()	EntropyNitrogenDioxide()
SpecificHeatNitrogenMonoxide()	EnthalpyNitrogenMonoxide()	EntropyNitrogenMonoxide()
SpecificHeatNonane()	EnthalpyNonane()	EntropyNonane()
SpecificHeatOctane()	EnthalpyOctane()	EntropyOctane()
SpecificHeatOxygen()	EnthalpyOxygen()	EntropyOxygen()
SpecificHeatPentane()	EnthalpyPentane()	EntropyPentane()
SpecificHeatPropane()	EnthalpyPropane()	EntropyPropane()
SpecificHeatWater()	EnthalpyWater()	EntropyWater()

Appendix A: A Microscopic Perspective on Availability and Irreversibility

A paper presented at the 1990 ASME Winter Annual Meeting, Houston, Texas.

Abstract

A thorough understanding of the Second Law of Thermodynamics is essential for the engineer working with energy systems. How the Second Law is presented to engineering students can have a profound effect on their perception of and dependence on this fundamental principle. An alternative is described to the usual way of presenting the microscopic Second Law. This approach provides an intuitive argument separate from and supportive of the axiomatic statements of Clausius and Kelvin-Planck. The impression can then be made on the student that the Second Law describes the inherent behavior of matter and energy rather than just the functioning of heat engines.

Nomenclature
e = energy associated with a particular level
E = total energy
H = enthalpy
n = population
N = number of members in the system
p = pressure
Q = heat
U = internal energy
S = entropy
T = temperature
V = volume
W = work

Greek
ψ = availability
φ = irreversibility

Differentials
δ = incremental change
Δ = incremental transfer
d = exact differential
đ = inexact differential

Subscripts
0 = depleted state
E = transport out of the system (exit)
J = associated with level "J"
I = transport into the system
S = property of the system

Introduction

Advances in computer technology have made possible greater levels of detail and pervasiveness of analysis of thermodynamic systems. What can be done or what is being done in research facilities is becoming what is expected in industry. Thermodynamic analysis of systems will be even more complex in the future as the variety of systems increases, as does the economic pressure to optimize their performance.

Rather than eliminating the need for a thorough understanding of principles by relegating the task of the engineer to that of a technician or computer operator, advancing technology requires ever greater understanding on the part of the engineer. Availability and irreversibility as thermodynamic concepts rather than waning into obscurity with the rise of the computer have become standard output items on the printouts generated by thermodynamic computer codes ranging from chemical process to steam power plant models.

A thorough understanding of what availability and irreversibility are and how these are a measure of the state of a system are a must for the engineer working in the area of energy systems. The way in which these concepts are presented and the emphasis which is placed on them can have a profound impact on whether the student later considers them familiar and even indispensable principles or relatively unimportant. A microscopic perspective is essential to a proper understanding of macroscopic thermodynamic systems–especially as those systems become more complex and less resembling simple heat engines.

Many students come away from a course on macroscopic thermodynamics with the impression that the Second Law only describes the limitations of heat engines rather than the manifestation of behaviors inherent within matter and energy. Students taking an additional course on microscopic thermodynamics may not have any more understanding–especially if the text opens with statistical mechanics. Statistical mechanics is often so formidable in and of itself that the purpose for its application to microscopic thermodynamics can be lost on the student.

An alternative is offered here to the typical presentation of microscopic thermodynamics. The Second Law and the associated quantities availability and irreversibility are developed from first principles from a microscopic perspective apart from the classical dependence on heat engines and apart from an emphasis on statistical mechanics. The correspondence between the microscopic and macroscopic expressions is then shown. Statistical mechanics could then be introduced as a means by which to characterize the particles too small to see and too numerous to count which comprise energy systems.

An Alternative Approach

In a typical course on classical thermodynamics, the Second Law and the associated quantities availability and irreversibility are developed by introducing the concept of heat engines and the axiomatic statements of Clausius and Kelvin-Planck (e.g. van Wylen and Sonntag, Chapters 5-7). A typical course on microscopic thermodynamics might introduce the concepts of probability and statistical measures of state and then illustrate these for the case of an ideal gas (e.g., Pierce, Chapters 5-8). In a somewhat novel approach, Holman inserts a chapter on statistical mechanics between a classical First Law chapter and a classical Second Law one.

In teaching thermodynamics, it is important to communicate that there is a relationship between the microscopic and the macroscopic. The hypothesis that is assumed, but may not be sufficiently emphasized for the student is that macroscopic phenomena are manifestations of microscopic ones. Thus, in order to fully understand the macroscopic one must study the microscopic as well.

Furthermore, the development of the Second Law along classical lines involving heat engines may give the erroneous impression that it applies only to heat engines–unless the presentation includes the microscopic perspective. Unfortunately, even if the microscopic perspective is presented, it may be difficult to see that there is a correspondence. Perhaps the greatest obstacle is the massive subject of statistical mechanics that typically composes the bulk of the text between particle kinetics and macroscopic properties of a system. Although unquestionably essential, statistical mechanics may be a stumbling block that regrettably divorces the microscopic and macroscopic perspectives.

The alternative strategy presented here for consideration is to take particle kinetics as far as possible toward classical thermodynamics without considering statistical mechanics, then show that there is a correspondence between the two before digressing into statistical mechanics. This strategy emphasizes the general concept of the correspondence first and the mathematical means of bridging the gap second. The goal of this strategy is to prevent the correspondence from being lost in the formidable details. The first step in developing these concepts is to build a consistent foundation of thermodynamic definitions and relationships that can be applied to either microscopic or macroscopic perspectives.

Thermodynamic Systems

A thermodynamic system is an abstract conceptual tool. A thermodynamic system is a closed region in space, the extent of which is defined by the system boundary. A system boundary does not itself occupy any space and has no physical existence. A system boundary can be defined anywhere one chooses, whether realistic or not, practical or not, physically constructible or not. One is not limited as to where a system boundary can be defined; however, the choice of a system boundary greatly affects its usefulness in any analyses.

The system boundary conceptually separates the system from its surroundings. The intersection of a system and its surroundings is the null set. The union of a system and its surroundings is the entire set (viz., the cosmos). The surroundings can be divided into the immediate surroundings and the ultimate surroundings; but this distinction is somewhat artificial and must always be arbitrarily limited in some way.

There are three types of systems. An isolated system is one in which there is no transfer of energy or matter with the surroundings. A closed system is one in which there is transfer of energy, but not matter. An open system is one in which there is an transfer of both energy and matter. It is essential to emphasize that the type of system is determined solely by how one defines the system boundary (van Wylen and Sonntag, pp. 17-19).

Energy Transfer Between a System and Its Surroundings

There are three modes by which energy is transferred between a system and its surroundings: transport, work, and heat. Transport refers to the transfer of energy by virtue of the transfer of matter. Transport applies only to open systems. Work refers to the transfer of energy by the application of a force (e.g., electromagnetic or gravitational). Heat refers to all other means of energy transfer (e.g., radiation). Work and heat apply to closed and open systems.

It is important to emphasize that work and heat are only modes by which energy is transferred. Energy is a property; whereas work and heat are not properties. A system and its surroundings can contain energy; whereas they cannot contain heat or work. Heat and work only exist at a system boundary and only exist while the transfer is taking place. The sign convention used by van Wylen and Sonntag (p. 82) and Pierce (p. 186) is positive for work done by the system on the surroundings and for heat transferred from the surroundings into the system.

The First Law of Thermodynamics

The concept of energy is inseparably linked to the First Law of Thermodynamics. The First Law is deduced from observations and can be stated as follows: there is a property (energy) that is conserved or remains constant in any process or interaction.

For an isolated system there is no transfer of energy with the surroundings. Therefore, the energy of an isolated system is constant. Denoting a change in the energy of the system by δE_S, this relationship can be expressed by Equation A.1.

$$\delta E_S = 0 \qquad (A.1)$$

For a closed system the only means of energy transfer are work and heat. Denoting an incremental transfer of heat by ΔQ and work by ΔW, this relationship can be expressed by Equation A.2 (van Wylen and Sonntag, p. 96).

$$\delta E_S - \Delta Q + \Delta W = 0 \qquad (A.2)$$

For an open system energy is transferred by means of transport, work, and heat. Denoting an incremental transport into and out of the system by δE_I and δE_E respectively, this relationship can be expressed by Equation A.3 (van Wylen and Sonntag, p. 124).

$$\delta E_S - \Delta Q + \Delta W - \delta E_I + \delta E_E = 0 \qquad (A.3)$$

Availability and Irreversibility

Energy is often incorrectly defined as the potential for doing work; whereas, availability more nearly fits this definition. More specifically, availability is the maximum energy which can be transferred between a system and its surroundings in such a way as to theoretically extract work (van Wylen and Sonntag, p. 282). Whether or not such a maximum extraction of work is possible introduces the concept of irreversibility. Irreversibility is defined as the difference between the ideal maximal and actual work extraction for a process. Denoting a change in availability by $\Delta \psi$ and irreversibility by $\Delta \varphi$, this relationship can be expressed by Equation A.4 (van Wylen and Sonntag, p. 276).

$$\Delta\varphi = -\Delta\psi - \Delta W \geq 0 \qquad (A.4)$$

The negative signs in the above expression arise from the sign convention for work being opposite of that for transport and heat transfer.

Because availability is the theoretical maximum, $-\Delta\psi$ must always be greater than or equal to ΔW. Thus, $\Delta\varphi$ must always be greater than or equal to zero. It is important to emphasize that this inequality which relates irreversibility to availability and work is not an empirically deduced principle–as is the First Law–it is a definition. It will be shown subsequently that this expression for a change in irreversibility, $\Delta\varphi$, is actually the Second Law. Expressions for the work, ΔW, are straightforward. What remains to be developed are expressions for the change in availability, $\Delta\psi$.

Energy is a property of a system, and as such depends only on the state of the system. Availability is based on ideal maximal work, which implies an interaction with the surroundings. Thus availability depends on the state of the system and its surroundings. It is intuitive that both measures are necessary; because there are many systems which contain significant energy, but which seem incapable of doing any work. It is also intuitive that the availability of a system depends on its relation to the surroundings. For example, a tank containing air at room temperature and 1 bar has little or no availability. However, if the atmospheric pressure were only 0.5 bar, the same tank of air–all other things being equal–would have availability to, for instance, drive a turbine.

Partitioning of Energy within a System

Central to the understanding of availability is the concept of energy partitioning; or how the energy is distributed within a system. In a system there are a finite multitude of attainable energy levels; and within each energy level there can be a number of states that have the same energy (Pierce, p. 157). Not all energy levels are attainable by a system. For instance, a level that exceeds the energy of the entire system would not be attainable. Furthermore, not all energy states within a level are attainable. Although the population of a level may be large, the Pauli Exclusion Principle asserts that the population of a unique energy state is limited to no more than one (Pierce, p. 170). Quantum theory asserts that the attainable energy states and thus levels can assume only discrete values (Pierce, p. 166).

Consider an isolated system containing N irreducible, distinct, although not necessarily distinguishable, members (N need not be large). The system has attainable energy levels, e_J. The population of each level is n_J. The populations, n_J, need not be large (typically many of the attainable energy levels in a system are empty). Two of the constraints on the system are the conservation of members;

$$\Sigma n_J = N \qquad (A.5)$$

and the conservation of total system energy (Pierce, p. 136).

$$\Sigma n_J e_J = E_S \quad (A.6)$$

For example, consider a system containing 3 members (abc) and having 5 discrete attainable energy levels (0-4) and a total system energy of 4, there are 15 different attainable partitionings (A-O) as illustrated in Table 1 (Pierce, p. 129).

Changes in Population and Attainable Energy Levels

partitioning	\multicolumn{5}{c}{energy levels}				
	0	1	2	3	4
A	ab				c
B	ac				b
C	bc				a
D	a	b	c		
E	b	a	c		
F	a	c	b		
G	c	a	b		
H	b	c	a		
I	c	b	a		
J	a		bc		
K	b		ac		
L	c		ab		
M		ab	c		
N		ac	b		
O		bc	a		

Consider an arbitrary change in the total energy of a system, δE_S, with populations and levels as indicated by Equation A.6. This change in energy is given by:

$$\delta E_S = \Sigma e_J \delta n_J + \Sigma n_J \delta e_J \quad (A.7)$$

The first term, $\Sigma e_J \delta n_J$, represents a change in energy due to a redistribution of populations. The second term, $\Sigma n_J \delta e_J$, represents a change in energy due to a change in energy levels. These changes are illustrated in Figures A.1 and A.2 respectively (Pierce, pp. 184-186).

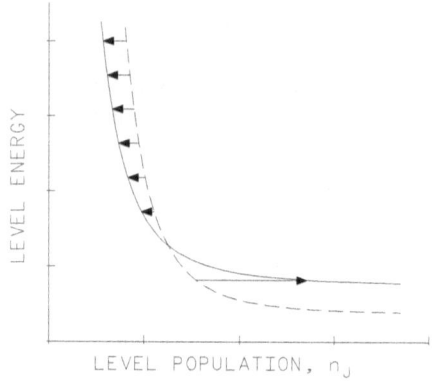

Figure A1. Change in Level Populations

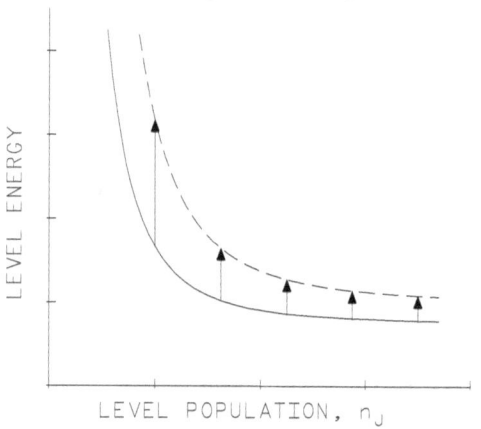

Figure A2. Change in Energy Levels

An analogy that may be helpful in understanding the changes illustrated in Figures A.1 and A.2 is that of parallel dams on parallel rivers, each with a different elevation (not series dams on a single river). In this illustration energy level is analogous to elevation and population is analogous to volume. The change in Figure A.1 is analogous to exchanging sediment from the upper reservoirs for water from the lower such that the elevations and the total volume of water remain constant. As a result of this change there is more volume in the upper reservoirs and less in the lower. The change in Figure A.2 is analogous to increasing the total sediment in the reservoirs such that the elevations increase, but the total volume of water remains constant. In both cases the energy of the system is increased by raising the center of gravity of a constant volume of water. However, in the first case, the attainable hydropower remains constant; whereas in the second case it increases.

The process in Figure A.1 is a redistribution of populations within the same levels; thus there is no useful work associated with this process. The process in Figure A.2 is a change in levels with the populations locked into each level. Useful work could be extracted to the surroundings as levels drop or input to the system as levels increase (Pierce, p. 185).

Availability and the Depleted State

As stated previously, expressions for the availability of a system must necessarily consider the state of the system and the state of the surroundings. For the purposes of thermodynamic analysis it is necessary to distinguish between the immediate and ultimate surroundings. Thus it is assumed that the immediate surroundings of a system can, at any given instant, be characterized in some way by energy levels and populations as implicitly assumed for the system. Furthermore, it is postulated that, at any given instant, there exists a "depleted state" of the system with respect to the immediate surroundings such that the system in that state will have zero availability (viz. unable to do any more useful work unless the state of the immediate surroundings changes).

This depleted state is designated by the subscript "0" (zero). The corresponding energy of the system is designated by E_{S0}, populations, n_{J0}, and levels, e_{J0}. This "depleted state" is not "absolute zero" or zero energy (i.e., e_{J0} are not necessarily zero). The "depleted state" is depleted relative to the immediate surroundings, not "absolutely depleted".

The difference between a change in energy and one in availability is called the "unavailable energy". Returning to the dam analogy, the "unavailable energy" is the energy associated with the volume below the turbine inlets. The "unavailable energy" for a dam is equal to the product of the elevation of the turbine inlet and the change in the corresponding volume of the reservoir. The microscopic equivalent of this "unavailable energy" is equal to the product of the depleted energy level and the change in the corresponding population or, $e_{J0}\delta n_J$. Subtracting this change in "unavailable energy" from the total change in energy (Equation A.7) yields Equation A.8 for the change in availability.

$$\Delta\psi = \Sigma(e_J - e_{J0})\delta n_J + \Sigma n_J \delta e_J \quad (A.8)$$

The first term in Equation A.8 accounts for the redistribution of populations and the second term for the change in energy levels from the present to the depleted state. Using the dam analogy, e_J and e_{J0} are the initial and final elevations respectively and n_J and n_{J0} the initial and final volumes respectively for reservoir "J". The first term is the elevation difference times the change in volume for each reservoir; and the second term is the volume times the change in elevation for each reservoir. Recognizing that $\delta e_{J0} = 0$, Equation A.9 can be differentiated to obtain Equation A.8. Therefore, Equation A.9 is the microscopic equation for the availability of a closed system.

$$\psi = \Sigma(e_J - e_{J0})n_J \quad (A.9)$$

Thus the availability of a closed system is equal to the sum of the product of the present populations and the difference between the present and depleted energy states. This expression also follows intuitively from the dam analogy. Equations A.3 and A.9 can be combined to give the microscopic equation for the availability of an open system.

$$\psi=\Sigma(e_{SJ}-e_{SJ0})n_{SJ}-\Sigma(e_{IJ}-e_{IJ0})n_{IJ}+\Sigma(e_{EJ}-e_{EJ0})n_{EJ} \quad (A.10)$$

Here the additional subscripts S, I, and E have been added to distinguish between the system, transport into, and out of (exiting) the system respectively. Consider the system described in Table 1 with an initial state of A and a depleted state of Z. In state Z consider the energy levels to have dropped from 01234 to 00112; and the populations to have changed from 20001 to 11100. The energy excess relative to the depleted state would be:

$$E-E_0=(2\times0+0\times1+0\times2+0\times3+1\times4)$$
$$-(1\times0+1\times0+1\times1+0\times1+0\times2)=3$$

The availability would be:

$$\psi_A=2\times(0-0)+0\times(1-0)+0\times(2-1)$$
$$+0\times(3-1)+1\times(4-2)=2$$

If the initial state had been D rather than A the excess energy would be the same; but the availability would be:

$$\psi_D=1\times(0-0)+1\times(1-0)+0\times(2-1)$$
$$+1\times(3-1)+0\times(4-2)=3$$

Similarly for initial states J and M respectively:

$$\psi_J=1\times(0-0)+0\times(1-0)+2\times(2-1)$$
$$+0\times(3-1)+0\times(4-2)=2$$

$$\psi_M=0\times(0-0)+2\times(1-0)+1\times(2-1)$$
$$+0\times(3-1)+0\times(4-2)=3$$

The microscopic expression for the change in irreversibility of a closed system can now be obtained by substituting Equation A.8 into Equation A.4 to yield Equation A.11.

$$\Delta\phi=-\Sigma(e_J-e_{J0})\delta n_J-\Sigma n_J\delta e_J-\Delta W\geq 0 \quad (A.11)$$

Similarly, the expression for an open system can be obtained in the same way using Equation A.10.

$$\Delta\phi=-\Sigma(e_{SJ}-e_{SJ0})\delta n_{SJ}-\Sigma n_{SJ}\delta e_{SJ}$$
$$+\Sigma(e_{IJ}-e_{IJ0})\delta n_{IJ}+\Sigma n_{IJ}\delta e_{IJ}$$
$$-\Sigma(e_{EJ}-e_{EJ0})\delta n_{EJ}-\Sigma n_{EJ}\delta e_{EJ}-\Delta W\geq 0 \quad (A.12)$$

Classical Thermodynamics

Equations A.11 and A.12, although not in their most familiar forms, are the Second Law of Thermodynamics for closed and open systems respectively. In developing these equations no mention has been made of pressure, temperature, entropy, or heat engines–all macroscopic concepts. It will now be shown that the above microscopic expressions have a direct correspondence to the classical macroscopic expressions.

Equation A.7 is the change in energy for a closed system. Only the second term, $\Sigma n_J \delta e_J$, contributed to the ideal work of the system. For a simple stationary classical system where only the internal energy, U, is significant and the only means of exchanging work with the surroundings is mechanical, the ideal work performed on the system is given by Equation A.13 in terms of the pressure, p, and the volume, V (van Wylen and Sonntag, p. 70).

$$-{\rm d}W_{IDEAL} = -pdV \tag{A.13}$$

Therefore, these two terms must be equal.

$$\Sigma n_J \delta e_J = -pdV \tag{A.14}$$

Here the distinction is made between exact differentials (i.e., properties of the system) and inexact differentials (i.e., work and heat) by the two symbols for differentiation, d and ${\rm d}$, respectively.

The classical expression of the First Law for a closed system that can be taken as one definition of entropy, S, and involving the temperature, T, is given by Equation A.15 (van Wylen and Sonntag, p. 212).

$$dU = TdS - pdV \tag{A.15}$$

For this system $dU = \delta E_S$; therefore, the second term in Equation A.7 must be equal to TdS.

$$\Sigma e_J \delta n_J = TdS \tag{A.16}$$

Furthermore, the second term in Equation A.8 must be equal to $T_0 dS$.

$$\Sigma e_{J0} \delta n_J = T_0 dS \tag{A.17}$$

Which means that Equation A.11 must be equivalent to Equation A.18.

$$\varphi = -(T-T_0)dS + pdV - {\rm\_}W \geq 0 \tag{A.18}$$

Adding Equations A.2, A.15, and A.18 yields Equation A.19.

$$\varphi = T_0 dS - {\rm\_}Q \geq 0 \tag{A.19}$$

which is the classical expression of the Second Law for a closed system (van Wylen and Sonntag, p. 277). The differential availability can be found by adding Equations A.2 and A.4 and subtracting Equation A.19.

$$\psi = dU - T_0 dS \tag{A.20}$$

which is the classical expression for the availability of a closed system (van Wylen and Sonntag, p. 283). Similarly, Equation A.12 can be used to determine corresponding expressions for an open system.

$$\varphi = T_0 dS_S - T_0 dS_I + T_0 dS_E - {\rm\_}Q \geq 0 \tag{A.21}$$
$$\psi = (dU_S - T_0 dS_S) - (dH_I - T_0 dS_I) + (dH_E - T_0 dS_E) \tag{A.22}$$

Here in the case of an open system, enthalpy, H, is the appropriate measure of energy for the inlets and exits.

Summary

The preceding derivations show that microscopic expressions for availability, irreversibility, and the Second Law can be developed apart from references to heat engines. Thus, reinforcing the concept that these are not restricted to heat engines. This also reinforces the veracity of the Clausius and Kelvin-Planck axioms.

The similarity and correspondence of the microscopic and macroscopic relationships illustrates that there is an equivalence of the two perspectives and reinforces the hypothesis that macroscopic phenomena are manifestations of microscopic ones.

The impression is also given that if one were somehow able to characterize the energy of every member of a system, then one could compute classical thermodynamic properties from particle kinetics as illustrated for the system in Table 1. This is also a logical point at which to introduce statistical mechanics as a means to solve the logistical problem of systems composed of members too small to be seen and too numerous to be counted.

References

Holman, J. P. 1974, *Thermodynamics*, 2nd. Ed., McGraw-Hill, New York.

Pierce, F. J. 1968, *Microscopic Thermodynamics*, International Textbook, Scranton, PA.

van Wylen, G. J. and R. E. Sonntag 1973, *Fundamentals of Classical Thermodynamics*, 2nd Ed., John Wiley and Sons, New York.

Appendix B. van der Waals EOS Program

The following program uses the van der Waals equation of state to compute the properties of an ideal fluid (liquid and vapor) as well as generate a Mollier Chart.

```
#define _CRT_SECURE_NO_DEPRECATE
#include <conio.h>
#include <stdio.h>
#include <stdlib.h>
#include <float.h>
#define _USE_MATH_DEFINES
#include <math.h>

double Cp=2.5;
double Zc=3./8.;
double Ac=9./8.;
double Bc=1./3.;
double Ho=0.;
double So=0.;

double fPr(double Tr,double Vr) /* reduced pressure */
  {
  if(Vr<=Bc)
     return(-999.);
  if(Tr<=0.)
     return(-999.);
  return((Tr/(Vr-Bc)-Ac/Vr/Vr)/Zc);
  }

double fFr(double Tr,double Vr) /* fugacity coefficient
    */
  {
  double Pr,Z;
  if(Vr<=Bc)
     return(-999.);
  if(Tr<=0.)
     return(-999.);
  Pr=fPr(Tr,Vr);
  if(Pr<=0.)
     return(-999.);
  Z=Zc*Pr*Vr/Tr;
  return(exp(Z-1.-log(Z)+log(Vr/(Vr-Bc))-Ac/Vr/Tr));
  }

double fHr(double Tr,double Vr) /* residual enthalpy */
  {
  double Pr,Z;
  if(Vr<=Bc)
     return(-999.);
```

```
  if(Tr<=0.)
    return(-999.);
  Pr=fPr(Tr,Vr);
  if(Pr<=0.)
    return(-999.);
  Z=Zc*Pr*Vr/Tr;
  return(Tr*(1.-Z)+Ac/Vr);
}

double HofTV(double Tr,double Vr) /* reduced enthalpy */
{
  return(Ho+Cp*Tr-fHr(Tr,Vr));
}

double fSr(double Tr,double Vr) /* residual entropy */
{
  double F,Pr,Hr;
  if(Vr<=Bc)
    return(-999.);
  if(Tr<=0.)
    return(-999.);
  Pr=fPr(Tr,Vr);
  if(Pr<=0.)
    return(-999.);
  F=fFr(Tr,Vr);
  if(F<=0.)
    return(-999.);
  Hr=fHr(Tr,Vr);
  return(Hr/Tr+log(F));
}

double SofTV(double Tr,double Vr) /* reduced entropy */
{
  double Pr;
  Pr=fPr(Tr,Vr);
  if(Pr<=0.)
    return(-999.);
  return(So+Cp*log(Tr)-fSr(Tr,Vr)-log(Pr));
}

typedef struct{double r,i;}COMPLEX;

int Cubic(double*c,COMPLEX*r)/* c0+c1*x+c2*x^2+x^3=0 */
{
  double a,arm,b,ca,ca3,cb,cb3,d,rm,t,x;
  memset(r,0,3*sizeof(COMPLEX));
  a=c[1]-c[2]*c[2]/3.;
  b=(2.*c[2]*c[2]*c[2]-9.*c[2]*c[1]+27.*c[0])/27.;
  d=b*b/4.+a*a*a/27.;
```

```
   if(d>=0.)
     {
     d=sqrt(d);
     ca3=-b/2.+d;
     cb3=-b/2.-d;
     if(ca3<0.)
        ca=-pow(-ca3,1./3.);
     else if(ca3>0.)
        ca=pow(ca3,1./3.);
     else
        ca=0.;
     if(cb3<0.)
        cb=-pow(-cb3,1./3.);
     else if(cb3>0.)
        cb=pow(cb3,1./3.);
     else
        cb=0.;
     r[0].r=ca+cb-c[2]/3.;
     r[1].r=r[2].r=-(ca+cb)/2.-c[2]/3.;
     r[1].i=(ca-cb)*sqrt(3.)/2.;
     r[2].i=-r[1].i;
     return(1);
     }
   rm=2.*sqrt(-a/3.);
   arm=a*rm;
   if(fabs(arm)>0.)
     x=3.*b/arm;
   else
     x=0.;
   t=acos(x)/3.;
   r[0].r=rm*cos(t)-c[2]/3.;
   r[1].r=rm*cos(t+2.*M_PI/3.)-c[2]/3.;
   r[2].r=rm*cos(t+4.*M_PI/3.)-c[2]/3.;
   return(3);
   }

void SatProps(double Tr,double*Pr,double*Vf,double*Vg)
   {
   int i;
   double c[3],P1,P2;
   COMPLEX r[3];
   if(Tr>=1.)
     {
     *Pr=*Vf=*Vg=1.;
     return;
     }
   P1=0.00001;
   P2=1.;
   for(i=0;i<32;i++)
```

```
      {
      *Pr=sqrt(P1*P2);
      c[0]=-27.*Pr[0]*Pr[0]/512./Tr/Tr/Tr;
      c[1]=27.*Pr[0]/64./Tr/Tr;
      c[2]=-(Pr[0]/8./Tr+1.);
      if(Cubic(c,r)!=3)
        {
        if(r[0].r>0.5)
          P1=*Pr;
        else
          P2=*Pr;
        continue;
        }
      *Vf=min(r[0].r,min(r[1].r,r[2].r))*Tr/Pr[0]/Zc;
      *Vg=max(r[0].r,max(r[1].r,r[2].r))*Tr/Pr[0]/Zc;
      if(fFr(Tr,*Vf)>fFr(Tr,*Vg))
        P1=*Pr;
      else
        P2=*Pr;
      }
  }

void SaturationTable()
  {
  double Hf,Hg,Ps,Sf,Sg,Ts,Vf,Vg,Zg;
  printf("van der Waals Equation of State\n");
  printf("saturation properties\n");
  printf("Trs     Prs     Vrf     Vrg     Zg     Hrf     Hrg
    Srf    Srg\n");
  for(Ts=0.5;Ts<1.001;Ts+=0.01)
    {
    SatProps(Ts,&Ps,&Vf,&Vg);
    Zg=Zc*Ps*Vg/Ts;
    Hf=HofTV(Ts,Vf);
    Hg=HofTV(Ts,Vg);
    Sf=SofTV(Ts,Vf);
    Sg=SofTV(Ts,Vg);
    if(Ho==0.)
      {
      Ho=-Hf;
      So=-Sf;
      Hf+=Ho;
      Hg+=Ho;
      Sf+=So;
      Sg+=So;
      }
    printf("%4.2lf %6.4lf %6.4lf %7.4lf %6.4lf %6.4lf
    %6.4lf %6.4lf %6.4lf\n",Ts,Ps,Vf,Vg,Zg,Hf,Hg,Sf,Sg);
    }
```

```
  }
#define n 20
#define m 25

void MollierChart()
  {
  int i,j;
  double Hr,Ps[m],Sr,Tr,Ts[m],Vf[m],Vg[m],Vr;
  FILE*fp;

  printf("creating Mollier chart\n");
  fp=fopen("Mollier.csv","wt");

/* subcritical isotherms */

  for(j=0;j<m;j++)
    {
    Tr=0.5+j*0.5/m;
    Ts[j]=Tr;
    SatProps(Tr,Ps+j,Vf+j,Vg+j);
    for(i=0;i<n;i++)
      {
      Vr=Bc*exp((i+1)*log(Vf[j]/Bc)/n);
      Sr=SofTV(Tr,Vr);
      Hr=HofTV(Tr,Vr);
      if(Hr<1.+0.65*Sr)
        fprintf(fp,"%1G,%1G\n",Sr,Hr);
      }

    fprintf(fp,"%1G,%1G\n",SofTV(Tr,Vf[j]),HofTV(Tr,Vf[j]
));

    fprintf(fp,"%1G,%1G\n",SofTV(Tr,Vg[j]),HofTV(Tr,Vg[j]
));
    for(i=0;i<n;i++)
      {
      Vr=Vg[j]*exp((i+1)*log(100./Vg[j])/n);
      Sr=SofTV(Tr,Vr);
      Hr=HofTV(Tr,Vr);
      fprintf(fp,"%1G,%1G\n",Sr,Hr);
      }
    fprintf(fp,"\n");
    }
/* vapor dome */

  for(j=0;j<m;j++)
```

```
      fprintf(fp,"%1G,%1G\n",SofTV(Ts[j],Vf[j]),HofTV(Ts[j]
      ,Vf[j]));
      fprintf(fp,"%1G,%1G\n",SofTV(Tr,1.),HofTV(Tr,1.));
      for(j=m-1;j>=0;j--)

        fprintf(fp,"%1G,%1G\n",SofTV(Ts[j],Vg[j]),HofTV(Ts[j]
        ,Vg[j]));
      fprintf(fp,"\n");

/* supercritical isotherms */

      for(Tr=1.;Tr<1.61;Tr+=0.02)
        {
        for(i=0;i<n;i++)
          {
          Vr=Bc*exp((i+1)*log(100./Bc)/n);
          Sr=SofTV(Tr,Vr);
          Hr=HofTV(Tr,Vr);
          if(Hr<1.+0.65*Sr)
             fprintf(fp,"%1G,%1G\n",Sr,Hr);
          }
        fprintf(fp,"\n");
        }

      fclose(fp);
      }

int main(int argc,char**argv,char**envp)
{
SaturationTable();
MollierChart();
return(0);
}
```

In addition to the file mollier.csv, the output of this program is as follows:

```
van der Waals Equation of State
saturation properties
Trs    Prs     Vrf     Vrg      Zg      Hrf     Hrg     Srf     Srg
0.50   0.0278  0.4068  45.9838  0.9584  0.0000  3.2163  0.0000  6.4326
0.51   0.0317  0.4091  40.8908  0.9543  0.0312  3.2357  0.0605  6.3440
0.52   0.0361  0.4114  36.5169  0.9500  0.0626  3.2547  0.1203  6.2590
0.53   0.0408  0.4138  32.7409  0.9455  0.0943  3.2733  0.1793  6.1774
0.54   0.0460  0.4163  29.4651  0.9407  0.1264  3.2914  0.2377  6.0988
0.55   0.0516  0.4188  26.6099  0.9358  0.1587  3.3090  0.2954  6.0232
0.56   0.0576  0.4214  24.1103  0.9307  0.1913  3.3261  0.3524  5.9503
0.57   0.0642  0.4241  21.9131  0.9254  0.2242  3.3427  0.4089  5.8799
0.58   0.0712  0.4269  19.9738  0.9199  0.2575  3.3588  0.4648  5.8118
0.59   0.0788  0.4297  18.2559  0.9142  0.2912  3.3743  0.5202  5.7459
0.60   0.0869  0.4326  16.7285  0.9082  0.3252  3.3893  0.5752  5.6820
0.61   0.0955  0.4356  15.3660  0.9022  0.3595  3.4037  0.6297  5.6201
```

```
0.62  0.1047  0.4387  14.1466  0.8959  0.3943  3.4175  0.6838  5.5599
0.63  0.1145  0.4419  13.0519  0.8894  0.4295  3.4307  0.7375  5.5013
0.64  0.1249  0.4451  12.0663  0.8827  0.4651  3.4433  0.7908  5.4442
0.65  0.1358  0.4485  11.1763  0.8759  0.5011  3.4552  0.8438  5.3886
0.66  0.1475  0.4520  10.3705  0.8688  0.5376  3.4665  0.8966  5.3343
0.67  0.1597  0.4556   9.6390  0.8616  0.5746  3.4771  0.9491  5.2812
0.68  0.1726  0.4593   8.9734  0.8542  0.6121  3.4871  1.0013  5.2292
0.69  0.1862  0.4632   8.3663  0.8466  0.6501  3.4963  1.0534  5.1782
0.70  0.2005  0.4672   7.8111  0.8388  0.6887  3.5047  1.1053  5.1282
0.71  0.2154  0.4713   7.3025  0.8308  0.7278  3.5124  1.1571  5.0790
0.72  0.2311  0.4756   6.8354  0.8227  0.7676  3.5193  1.2088  5.0306
0.73  0.2475  0.4801   6.4055  0.8143  0.8080  3.5254  1.2605  4.9829
0.74  0.2646  0.4848   6.0092  0.8057  0.8490  3.5306  1.3121  4.9358
0.75  0.2825  0.4896   5.6431  0.7970  0.8908  3.5349  1.3638  4.8893
0.76  0.3011  0.4947   5.3041  0.7880  0.9333  3.5383  1.4155  4.8432
0.77  0.3205  0.5000   4.9898  0.7788  0.9766  3.5408  1.4674  4.7975
0.78  0.3406  0.5055   4.6978  0.7694  1.0207  3.5422  1.5194  4.7521
0.79  0.3616  0.5113   4.4259  0.7597  1.0657  3.5425  1.5716  4.7069
0.80  0.3834  0.5174   4.1725  0.7498  1.1117  3.5418  1.6242  4.6618
0.81  0.4059  0.5238   3.9357  0.7396  1.1586  3.5398  1.6770  4.6168
0.82  0.4293  0.5306   3.7141  0.7292  1.2067  3.5366  1.7303  4.5717
0.83  0.4535  0.5377   3.5064  0.7185  1.2559  3.5321  1.7841  4.5265
0.84  0.4786  0.5453   3.3113  0.7075  1.3064  3.5261  1.8384  4.4810
0.85  0.5045  0.5534   3.1276  0.6961  1.3582  3.5186  1.8935  4.4351
0.86  0.5312  0.5620   2.9545  0.6844  1.4116  3.5094  1.9494  4.3887
0.87  0.5589  0.5712   2.7909  0.6723  1.4666  3.4984  2.0062  4.3416
0.88  0.5874  0.5811   2.6360  0.6598  1.5234  3.4854  2.0641  4.2936
0.89  0.6167  0.5918   2.4889  0.6468  1.5823  3.4702  2.1233  4.2445
0.90  0.6470  0.6034   2.3488  0.6332  1.6435  3.4525  2.1842  4.1941
0.91  0.6782  0.6161   2.2151  0.6190  1.7074  3.4320  2.2468  4.1420
0.92  0.7102  0.6302   2.0869  0.6041  1.7743  3.4083  2.3118  4.0879
0.93  0.7432  0.6459   1.9634  0.5884  1.8449  3.3808  2.3796  4.0311
0.94  0.7771  0.6637   1.8438  0.5716  1.9199  3.3487  2.4509  3.9709
0.95  0.8119  0.6841   1.7271  0.5535  2.0004  3.3110  2.5268  3.9063
0.96  0.8476  0.7082   1.6118  0.5337  2.0881  3.2659  2.6088  3.8357
0.97  0.8843  0.7376   1.4960  0.5114  2.1858  3.2107  2.6998  3.7563
0.98  0.9219  0.7755   1.3761  0.4855  2.2991  3.1398  2.8050  3.6629
0.99  0.9605  0.8309   1.2430  0.4522  2.4419  3.0391  2.9382  3.5415
1.00  1.0000  1.0000   1.0000  0.3750  2.7616  2.7616  3.2458  3.2458
creating Mollier chart
```

Appendix C. List of Refrigerants and Properties

Thermodynamic properties for the following fluids are included in the vapor-compression refrigeration cycle spreadsheet (refrigerants.xls):

Name	Formula	MW 1/mole	Tc °K	Pc kPa	Vc cm³/gm	Zc -	R J/kg/°K	Pitzer acentric	Cpl kJ/kg/°C	Cpv kJ/kg/°C
R11	CCl3F	137.37	471.2	4409	1.806	0.2792	60.5	0.1915	0.847	0.476
R12	CF2Cl2	120.93	385.2	4115	1.792	0.2785	68.8	0.1764	0.856	0.438
R13	CClF3	104.46	301.5	3870	1.730	0.2790	79.6	0.1792	0.888	0.442
R13b1	CBrF3	148.91	340.2	3964	1.343	0.2802	55.8	0.1722	0.607	0.335
R14	CF4	88.01	227.7	3745	1.598	0.2782	94.5	0.1740	0.837	0.412
R21	CHCl2F	102.92	451.6	5167	1.848	0.2616	80.8	0.2038	0.923	0.488
R22	CHClF2	86.48	369.2	4977	1.906	0.2672	96.1	0.2220	0.940	0.485
R23	CHF3	70.01	299.1	4836	1.904	0.2593	118.8	0.2654	1.323	0.498
R40	CH3Cl	50.49	416.3	6759	2.793	0.2754	164.7	0.1634	1.452	0.628
R50	CH4	16.04	190.7	4641	6.181	0.2903	518.4	0.0112	3.433	1.758
R113	Cl2FC-CClF2	187.38	487.2	3392	1.786	0.2802	44.4	0.2518	0.842	0.603
R114	ClF2C-CClF2	170.92	418.8	3257	1.724	0.2756	48.6	0.2544	0.758	0.524
R115	F3C-CClF2	154.47	353.1	3129	1.627	0.2678	53.8	0.2494	0.883	0.532
R134a	CH2FCF3	102.03	374.2	4059	1.953	0.2600	81.5	0.3270	1.259	0.632
R142b	CH3CClF2	100.50	410.3	4246	2.299	0.2876	82.7	0.2501	0.603	0.636
R152a	CHF2CH3	66.05	386.4	4520	2.717	0.2525	125.9	0.2694	0.126	0.762
R170	C2H6	30.07	305.4	4894	5.181	0.3003	276.5	0.0988	2.261	1.256
R290	C3H8	44.10	369.9	4251	4.536	0.2764	188.6	0.1527	2.188	1.181
R500	R12/152a	99.30	375.3	4173	2.012	0.2672	83.7	0.2170	0.904	0.561
R502	R22/115	111.60	355.3	4075	1.784	0.2746	74.5	0.2186	0.695	0.486
R503	R12/23	87.20	292.7	4359	1.773	0.2770	95.3	0.1978	0.955	0.444
R504	R32/115	79.20	335.3	4439	1.818	0.2293	105.0	0.2650	1.055	0.553
R505	R12/31	103.50	390.9	4727	1.863	0.2804	80.3	0.1776	0.800	0.465
R506	R12/31	93.70	414.8	5167	1.815	0.2547	88.7	0.2345	0.867	0.557
R600	C4H10	58.10	425.2	3797	4.382	0.2735	143.1	0.2333	2.010	1.340
R600a	C4H10	58.10	408.1	3648	4.520	0.2823	143.1	0.1848	2.177	1.424
R702	H2	2.02	33.0	1293	64.140	0.6097	4124.2	-0.2184	13.198	18.782
R704	He	4.00	5.2	248	14.360	0.3297	2077.6	-0.3486	1.739	3.478
R717	NH3	17.03	405.4	11330	4.252	0.2434	488.2	0.2558	4.690	1.870
R718	H2O	18.02	647.1	22064	3.106	0.2294	461.5	0.3444	4.217	1.837
R720	Ne	20.18	44.4	2654	2.070	0.3003	412.0	-0.0373	1.908	0.811
R728	N2	28.01	126.2	3394	3.215	0.2913	296.8	0.0402	2.063	0.923
R729	Air	28.97	132.4	3774	3.047	0.3026	287.0	0.0225	1.785	0.788
R732	O2	32.00	154.8	5080	2.294	0.2899	259.8	0.0229	1.665	0.907
R740	Air	39.95	150.9	4898	1.867	0.2913	208.1	-0.0039	1.150	0.317
R744	CO2	44.01	304.2	7392	2.156	0.2773	188.9	0.2280	1.983	0.441
R1150	CH2=CH2	28.05	283.1	5117	4.369	0.2664	296.4	0.0787	2.373	1.187
R1270	CH3CH=CH2	42.10	364.9	4621	4.540	0.2911	197.5	0.1445	2.462	0.976

Appendix D. Approximate Properties for Ideal Gases

The constant pressure and constant volume specific heats were introduced in Chapter 10 (Equations 10.1 and 10.2). These are equal to the partial derivatives of enthalpy at constant pressure and of internal energy at constant volume, respectively. For an ideal gas, that is, one that obeys the ideal gas equation of state, $pv=RT$, these are independent of pressure and volume, making them functions of temperature alone. The enthalpy of an ideal gas is then:

$$h = \int C_p dT \tag{D.1}$$

Also for an ideal gas, we can substitute $dh=CpdT$ and $v=RT/p$ into Equation 7.9 to obtain:

$$C_p dT = Tds + RT \frac{dp}{p} \tag{D.2}$$

We then solve Equation D.2 for ds and integrate to obtain:

$$s = \int C_p \frac{dT}{T} - R \int \frac{dp}{p} \tag{D.3}$$

If Cp is constant, Equations D.1 and D.3 simplify to:

$$h - h_{REF} = C_p (T - T_{REF}) \tag{D.4}$$

The choice of h_{REF}, s_{REF}, p_{REF}, and T_{REF} are arbitrary. A natural choice is zero for the first two, 25°C/77°F, and 1 atmosphere. This is the reference used in the Otto cycle spreadsheet. Returning to Equation D.2 we see that for an isentropic process:

$$C_p dT = RT \frac{dp}{p} \tag{D.5}$$

This equation can be rearranged and integrated:

$$\int_{T_1}^{T_2} C_p \frac{dT}{T} = R \int_{p_1}^{p_2} \frac{dp}{p} \tag{D.6}$$

If Cp is constant this can be solved to obtain:

$$\frac{T_2}{T_1} = \left(\frac{p_2}{p_1}\right)^{\left(\frac{R}{C_p}\right)} \tag{D.7}$$

The ratio of Cp/Cv is given the symbol k and is called the *isentropic exponent*. We note that $h=u+pv$, so that for an ideal gas $Cp-Cv=R$. By substituting this definition of k, the exponent in Equation D.7 becomes $(k-1)/k$. Since, for an ideal gas, $p_1v_1/T_1=p_2v_2/T_2$, we can write for an isentropic process:

$$\frac{p_2}{p_1} = \left(\frac{v_1}{v_2}\right)^k \tag{D.8}$$

also by D. James Benton

3D Articulation: Using OpenGL, ISBN-9798596362480, Amazon, 2021 (book 3 in the 3D series).

3D Models in Motion Using OpenGL, ISBN-9798652987701, Amazon, 2020 (book 2 in the 3D series.

3D Rendering in Windows: How to display three-dimensional objects in Windows with and without OpenGL, ISBN-9781520339610, Amazon, 2016 (book 1 in the 3D series).

A Synergy of Short Stories: The whole may be greater than the sum of the parts, ISBN-9781520340319, Amazon, 2016.

Azeotropes: Behavior and Application, ISBN-9798609748997, Amazon, 2020.

bat-Elohim: Book 3 in the Little Star Trilogy, ISBN-9781686148682, Amazon, 2019.

Boilers: Performance and Testing, ISBN: 9798789062517, Amazon 2021.

Combined 3D Rendering Series: 3D Rendering in Windows®, 3D Models in Motion, and 3D Articulation, ISBN-9798484417032, Amazon, 2021.

Complex Variables: Practical Applications, ISBN-9781794250437, Amazon, 2019.

Compression & Encryption: Algorithms & Software, ISBN-9781081008826, Amazon, 2019.

Computational Fluid Dynamics: an Overview of Methods, ISBN-9781672393775, Amazon, 2019.

Computer Simulation of Power Systems: Programming Strategies and Practical Examples, ISBN-9781696218184, Amazon, 2019.

Contaminant Transport: A Numerical Approach, ISBN-9798461733216, Amazon, 2021.

CPUnleashed! Tapping Processor Speed, ISBN-9798421420361, Amazon, 2022.

Curve-Fitting: The Science and Art of Approximation, ISBN-9781520339542, Amazon, 2016.

Death by Tie: It was the best of ties. It was the worst of ties. It's what got him killed., ISBN-9798398745931, Amazon, 2023.

Differential Equations: Numerical Methods for Solving, ISBN-9781983004162, Amazon, 2018.

Equations of State: A Graphical Comparison, ISBN-9798843139520, Amazon, 2022.

Evaporative Cooling: The Science of Beating the Heat, ISBN-9781520913346, Amazon, 2017.

Forecasting: Extrapolation and Projection, ISBN-9798394019494, Amazon 2023.

Heat Engines: Thermodynamics, Cycles, & Performance Curves, ISBN-9798486886836, Amazon, 2021.

Heat Exchangers: Performance Prediction & Evaluation, ISBN-9781973589327, Amazon, 2017.

Heat Recovery Steam Generators: Thermal Design and Testing, ISBN-9781691029365, Amazon, 2019.

Heat Transfer: Heat Exchangers, Heat Recovery Steam Generators, & Cooling Towers, ISBN-9798487417831, Amazon, 2021.

Heat Transfer Examples: Practical Problems Solved, ISBN-9798390610763, Amazon, 2023.

The Kick-Start Murders: Visualize revenge, ISBN-9798759083375, Amazon, 2021.

Jamie2: Innocence is easily lost and cannot be restored, ISBN-9781520339375, Amazon, 2016-18.

Kyle Cooper Mysteries: Kick Start, Monte Carlo, and Waterfront Murders, ISBN-9798829365943, Amazon, 2022.

The Last Seraph: Sequel to Little Star, ISBN-9781726802253, Amazon, 2018.

Little Star: God doesn't do things the way we expect Him to. He's better than that! ISBN-9781520338903, Amazon, 2015-17.

Living Math: Seeing mathematics in every day life (and appreciating it more too), ISBN-9781520336992, Amazon, 2016.

Lost Cause: If only history could be changed..., ISBN-9781521173770, Amazon, 2017.

Mass Transfer: Diffusion & Convection, ISBN-9798702403106, Amazon, 2021.

Mill Town Destiny: The Hand of Providence brought them together to rescue the mill, the town, and each other, ISBN-9781520864679, Amazon, 2017.

Monte Carlo Murders: Who Killed Who and Why, ISBN-9798829341848, Amazon, 2022.

Monte Carlo Simulation: The Art of Random Process Characterization, ISBN-9781980577874, Amazon, 2018.

Nonlinear Equations: Numerical Methods for Solving, ISBN-9781717767318, Amazon, 2018.

Numerical Calculus: Differentiation and Integration, ISBN-9781980680901, Amazon, 2018.

Numerical Methods: Nonlinear Equations, Numerical Calculus, & Differential Equations, ISBN-9798486246845, Amazon, 2021.

Orthogonal Functions: The Many Uses of, ISBN-9781719876162, Amazon, 2018.

Overwhelming Evidence: A Pilgrimage, ISBN-9798515642211, Amazon, 2021.

Particle Tracking: Computational Strategies and Diverse Examples, ISBN-9781692512651, Amazon, 2019.

Plumes: Delineation & Transport, ISBN-9781702292771, Amazon, 2019.

Power Plant Performance Curves: for Testing and Dispatch, ISBN-9798640192698, Amazon, 2020.

Practical Linear Algebra: Principles & Software, ISBN-9798860910584, Amazon, 2023.
Props, Fans, & Pumps: Design & Performance, ISBN-9798645391195, Amazon, 2020.
Remediation: Contaminant Transport, Particle Tracking, & Plumes, ISBN-9798485651190, Amazon, 2021.
ROFL: Rolling on the Floor Laughing, ISBN-9781973300007, Amazon, 2017.
Seminole Rain: You don't choose destiny. It chooses you, ISBN-9798668502196, Amazon, 2020.
Septillionth: 1 in 10^{24}, ISBN-9798410762472, Amazon, 2022.
Software Development: Targeted Applications, ISBN-9798850653989, Amazon, 2023.
Software Recipes: Proven Tools, ISBN-9798815229556, Amazon, 2022.
Steam 2020: to 150 GPa and 6000 K, ISBN-9798634643830, Amazon, 2020.
Thermochemical Reactions: Numerical Solutions, ISBN-9781073417872, Amazon, 2019.
Thermodynamic and Transport Properties of Fluids, ISBN-9781092120845, Amazon, 2019.
Thermodynamic Cycles: Effective Modeling Strategies for Software Development, ISBN-9781070934372, Amazon, 2019.
Version-Independent Programming: Code Development Guidelines for the Windows® Operating System, ISBN-9781520339146, Amazon, 2016.
The Waterfront Murders: As you sow, so shall you reap, ISBN-9798611314500, Amazon, 2020.
Weather Data: Where To Get It and How To Process It, ISBN-9798868037894, Amazon, 2023.

www.ingramcontent.com/pod-product-compliance
Lightning Source LLC
Chambersburg PA
CBHW030848180526
45163CB00004B/1501